About IFPRI

The International Food Policy Research Institute (IFPRI®) was established in 1975 to identify and analyze alternative national and international strategies and policies for meeting food needs of the developing world on a sustainable basis, with particular emphasis on low-income countries and on the poorer groups in those countries. While the research effort is geared to the precise objective of contributing to the reduction of hunger and malnutrition, the factors involved are many and wide-ranging, requiring analysis of underlying processes and extending beyond a narrowly defined food sector. The Institute's research program reflects worldwide collaboration with governments and private and public institutions interested in increasing food production and improving the equity of its distribution. Research results are disseminated to policymakers, opinion formers, administrators, policy analysts, researchers, and others concerned with national and international food and agricultural policy.

IFPRI is a member of the CGIAR Consortium.

2011 GLOBAL FOOD POLICY REPORT

IFPRI®

INTERNATIONAL FOOD POLICY RESEARCH INSTITUTE

International Food Policy Research Institute
2033 K Street, NW
Washington, DC 20006-1002, USA
Telephone: +1-202-862-5600
www.ifpri.org

DOI: 10.2499/9780896295476

Library of Congress Cataloging-in-Publication Data

2011 global food policy report / International Food Policy Research Institute.
p. cm.
title: Global food policy report
Includes bibliographical references.
ISBN 978-0-89629-547-6 (alk. paper)
1. Nutrition policy. I. International Food Policy Research Institute. II. Title: Global food policy report.
TX359.A12 2012
363.8'62—dc23

2012010851

PHOTO CREDITS

Cover image: © 2011 Tim Dirven/Panos
Chapter images: Page x © 2011 Tim Dirven/Panos; page 14 © 2011 G.M.B. Akash/Panos; page 24 © 2011 Sven Torfinn/Panos; page 38 © 2011 Patrick Brown/Panos; page 48 © 2008 Warren Clarke/Panos; page 54 © 2010 Jenny Matthews/Panos; page 62 © 2011 Sven Torfinn/Panos; page 68 © 2011 Zerihun Sewunet/ILRI; page 78 © 2011 Fernando Moleres/Panos.

Cover design: Julia Vivalo / Book design and layout: David Popham.

Contents

Preface

Since 2007, two rounds of food price hikes have contributed to millions of people being hungry or malnourished. The same factors that contributed to the 2007–08 food price crisis triggered similar price spikes in 2011—factors including a declining growth rate of agricultural productivity, high energy prices leading to expanded biofuel production, depreciation of the US dollar, strong demand from emerging economies for agricultural products, and weather shocks. With such complex forces at work, it is clear that the food policies necessary to ensure that all people have access to safe, sufficient, nutritious, and sustainably grown food must go beyond traditional agricultural production. Fittingly, demand for evidence-based research to inform those policies is higher than ever, and the International Food Policy Research Institute (IFPRI) produces global public goods to respond to that need.

IFPRI's *2011 Global Food Policy Report*—the first in a new annual series—provides an in-depth, contextualized look at the past year's major food policy developments and events. It both raises and answers these key questions: What happened in food policy in 2011 and why? What challenges and opportunities resulted? What could have been done differently? What should be done in the future?

In 2011, agriculture moved to the forefront of the international development agenda. In addition to producing adequate food, agriculture's crucial role in improving nutrition and health, sustainably making use of land and other natural resources, and helping to address global threats like climate change has received long-overdue recognition. Investments in the sector are rising, and contributions are coming from industrialized countries as well as emerging and developing economies, the private sector, and philanthropic entities. In addition to higher investments, policymakers also scaled up collaboration across borders, in particular in their efforts to control food price volatility through the provision of better market information. This type of global policymaking must continue to take into account that legislation in one country (particularly trade and environmental policies regarding biofuels) can harm food security in others. International agenda-setting meetings, like the Rio+20 United Nations Conference on Sustainable Development in June 2012, cannot neglect the concerns of the poor. As the humanitarian crisis in the Horn of Africa starkly reminds us, however, research agendas and information-sharing are not enough to avert or solve a problem; preventive actions are also needed.

The topics covered in the *2011 Global Food Policy Report* were selected after numerous consultations by a strategic advisory council consisting of policymakers, researchers, and other experts that sought to represent the most profound, relevant, and broadly applicable food policy issues that arose in 2011. IFPRI's Board of Trustees and senior staff then provided feedback on major development and research topics, and a review of related print and broadcast media from 2011 was conducted. Finally, leading policymakers and food experts from around the world were asked for their opinions on how to best capture national and regional perspectives.

Contributions were commissioned from experts, scholars, and stakeholders on topics that represent either a new development in food policy, a major change in food policy, or a new way of looking at a food policy issue. The topics are regional or global in scope and feature high-quality research results as well as expert opinions that will enhance the quality of debate.

IFPRI's *2011 Global Food Policy Report* is the first of its kind, and I hope it will contribute to an enriched research agenda that informs sound food policies to the benefit of the world's poorest and most vulnerable people. I welcome your feedback, comments, and suggestions at ifpri@cgiar.org.

SHENGGEN FAN
Director General

Acknowledgments

This report was prepared under the overall leadership of Shenggen Fan and a core team comprising Alexander J. Stein, Zhenya Karelina, Klaus von Grebmer, Rajul Pandya-Lorch, and Gwendolyn Stansbury. It benefitted greatly from the strategic insights of a committee of advisers, including Robert Bos, Margaret Catley-Carlson, Marion Guillou, Monty Jones, Agnes M. Kalibata, Michiel A. Keyzer, Justin Y. Lin, Mari E. Pangestu, Martin Pineiro, Prabhu Pingali, Beatriz da Silveira Pinheiro, Keming Qian, M. S. Swaminathan, Eric Tollens, Rhoda Tumusiime, Joachim von Braun, Emorn Wasantwisut, and Derek Yach. IFPRI's Board of Trustees provided additional guidance and inputs, as did the members of IFPRI's Senior Management Team.

Excellent text and data contributions were made by Perrihan Al-Riffai, Kym Anderson, Suresh Babu, Ousmane Badiane, Nienke Beintema, Samuel Benin, Deborah Brautigam, Clemens Breisinger, Bruce Campbell, Rahul Chaturvedi, Kevin Chen, Cindy Cox, S. Mahendra Dev, Betina Dimaranan, Paul Dorosh, Olivier Ecker, Shenggen Fan, Heidi Fritschel, Sara Gustafson, Derek Headey, Jikun Huang, Kabba Joiner, P. K. Joshi, Suneetha Kadiyala, Zhenya Karelina, Jawoo Koo, David Laborde, Rachel Licker, Tsitsi Makombe, Sohail J. Malik, Paswel Marenya, Geraldo B. Martha Jr., John McDermott, Ruth Meinzen-Dick, Siwa Msangi, Gerald Nelson, Alejandro Nin Pratt, Ephraim Nkonya, Tolulope Olofinbiyi, Steven Were Omamo, Robert Paarlberg, Amanda Palazzo, Rajul Pandya-Lorch, Amber Peterman, Prabhu Pingali, Beatriz da Silveira Pinheiro, Nilam Prasai, Agnes Quisumbing, Jagdeesh Rao Puppala, Claudia Ringler, M. S. Swaminathan, Maria Theresa Tenorio, Peter Timmer, Maximo Torero, Klaus von Grebmer, Stanley Wood, Derek Yach, Sivan Yosef, and Bingxin Yu.

In addition, the following people provided thoughtful statements about what influenced food policy in 2011: Bekele Geleta, José Graziano da Silva, Marion Guillou, Jeremy Hobbs, Michiel A. Keyzer, Rachel Kyte, Jiayang Li, Justin Yifu Lin, David Nabarro, Kanayo Nwanze, John Parker, Carlos Perez del Castillo, Keming Qian, Rajiv Shah, Josette Sheeran, Kathy Spahn, Eric Tollens, Joachim von Braun, and Emorn Wasantwisut.

Production of this report was led by IFPRI's Publications Department, including Adrienne Chu, Patricia Fowlkes, Heidi Fritschel, Corinne Garber, Michael Go, Marcia MacNeil, Lucy McCoy, Andrea Pedolsky, David Popham, Ashley St. Thomas, Julia Vivalo, and John Whitehead. In addition, valuable research support was provided by Joanna Brzeska, Zhenya Karelina, Tolulope Olofinbiyi, and Ana Ramirez.

The report benefitted greatly from careful peer review by IFPRI's Publications Review Committee, chaired by Gershon Feder, and many anonymous scholars and experts who reviewed the research and provided insightful comments on the preliminary drafts.

OVERVIEW

Major Food Policy Developments in 2011

Shenggen Fan, IFPRI

THE YEAR 2011 HIGHLIGHTED ONGOING CHALLENGES to global food security, from food price volatility, extreme weather shocks, and famine to unrest and conflicts. On the policy front, major developments at the global and national levels both offered grounds for encouragement and pointed to areas where further action is needed.

First, the good news: after many years of neglect, agriculture and food security are back on the development and political agendas. Both China and India continued to expand their spending on food security and agricultural production. Some 20 African countries have adopted national agricultural and food security investment plans in which they will devote 10 percent of their national budget to agriculture to achieve agricultural growth of 6 percent a year. The US Agency for International Development (USAID) moved forward with its Feed the Future Initiative, begun in 2010, and the World Bank Group maintained its recent increased annual commitments to agriculture and related sectors at about US$6 billion. The Consultative Group on International Agricultural Research (CGIAR)—a global partnership for sustainable development, of which IFPRI is a part—initiated an array of large, innovative research programs in 2011. And the Bill & Melinda Gates Foundation refreshed its agriculture strategy with a strong focus on agricultural development in Sub-Saharan Africa and South Asia.

More broadly, agriculture was increasingly seen as part of a larger context. It is becoming clear that agriculture contributes not just to food production, but also to human nutrition and health—conditions that in turn can affect agricultural productivity and overall economic growth. Agriculture is also an important element in a number of other interlocking systems. It has strong ties to water, land, and energy, which are, like agriculture itself, under increasing pressure. And many of the events of 2011 underlined how food security—that is, availability of and access to sufficient, safe, nutritious food to maintain a healthy and active life—is linked to other notions of security. These include economic security (related to employment, incomes, and gender), sociopolitical security (related to inequality, governance, and conflicts), and environmental security (related to natural resources).

New thinking has been accompanied by new actors entering the global food system. In 2011, for the first time, the agriculture ministers of the Group of 20 (G20) countries met and agreed to work together to tackle food price volatility and food insecurity. Emerging economies such as Brazil, China, and India have gained an increasing voice in international decisionmaking, moving from being aid recipients to aid donors and trading partners, with their own global agendas.

This overview reviews the major food policy developments of 2011, drawing largely on the chapters in this report, which look back at the year in detail.

FOOD PRICE LEVELS AND VOLATILITY

Global food prices rose during the first half of 2011 and fell during the second half of the year. The food price index of the Food and Agriculture Organization of the United Nations, which measures monthly change in the international prices of a basket of food commodities, reached a record high in February but moved steadily downward from June to December, ending lower for the year. Still, food price volatility remained high in 2011.

The factors that pushed up prices during the 2007–08 food price crisis were again at play during the 2010–11 crisis, including high oil prices, biofuel policies that promote the expansion of biofuel production, increased weather-related shocks such as droughts and floods, and growing demand from emerging economies. Further, the world remains vulnerable to food price swings because grain reserves are extremely low and staple grains are exported by just a few countries. However, favorable harvests in major producing regions and a stronger US dollar induced a fall in dollar-denominated prices during the second half of the year.

What do rising or volatile food prices mean for the poor? Higher food prices cut into the budgets of poor consumers but could raise the incomes of poor producers if they produce more than they consume. Volatile food prices, however, harm both consumers and producers by increasing uncertainty and making it difficult for households to budget for food consumption and to plan for production. Still, more needs to be learned about the specific impacts of price volatility on the diets of the poor, particularly women and children. In Ethiopia, for example, research on the 2007–08 food price crisis found that female-headed households were especially vulnerable to food price shocks.[1]

Shifts in food prices stimulated new policies and initiatives during the year. As mentioned, the G20 ministers of agriculture came together to design an action plan to reduce price volatility, regulate commodity markets, and promote long-term agricultural productivity. Toward the end of the year, the countries of the Association of Southeast Asian Nations, plus China, Japan, and South Korea (altogether known as ASEAN+3) established an emergency rice reserve to help ensure long-term food security in the region.

Some national policies taken in response to changes in food prices may have increased the strain on the global food system. To raise producer incomes, the government of Thailand, the world's largest exporter of milled rice, established a rice subsidy scheme that threatened to shrink its exports and contribute to higher global rice prices—a trend observed in the second half of the year. Several countries, including China, turned to large grain imports to build up strategic reserves, raising concerns about tighter grain markets.

NATURAL AND HUMAN-CAUSED SHOCKS

The world saw some of the most severe natural disasters on record in 2011. The 9.0-magnitude earthquake and tsunami in Japan; the severe floods or storms in Brazil, Pakistan, the Philippines, Thailand, and the United States; and the drought in the Horn of Africa imposed large economic losses during the year. According to the International Disaster Database, more than 200 natural disasters, affecting nearly 100 million people around the world, occurred during the year.[2] Munich Re, a reinsurance company in Germany, estimated that 2011 natural disasters imposed economic losses of a record US$380 billion—more than double those of 2010 and far above the record losses of 2005.[3] Poor and hungry people are particularly susceptible to these natural shocks.

In the Horn of Africa, severe drought due to consecutive poor rainy seasons was the worst experienced in 60 years. Extreme drought conditions triggered a widespread crisis in the region that was especially catastrophic in Somalia. Many parts of the Horn, especially the lowland areas, saw large crop losses, significant depletion of grazing resources, skyrocketing food prices, and substantial livestock and human mortality. The dire situation attracted belated policy and media attention as more than 13 million people, principally pastoralists and farmers, were affected and their food and nutrition security was severely undermined. Vulnerable groups such as women and children experienced acute food insecurity and undernutrition. The United Nations Children's Fund reported that more than 320,000 children suffered from severe malnutrition at the height of the crisis.

Droughts in the Horn of Africa are not new, but the scale of the 2010–11 crisis has been unusual. Although exposure to natural shocks is inevitable, human vulnerability to these shocks is not. Reducing vulnerability means improving society's ability to cope and build resiliency in the face of future shocks. Given the severity of the drought in the Horn of Africa and the frequency of humanitarian emergencies in the region, a concerted effort is needed to catalyze a transformation, combining innovation, experimentation, and political commitment to enhance resiliency and mitigate the chronic stresses that also impede progress in the region.

CLIMATE CHANGE

The record-breaking extreme weather events of 2011 suggested that climate change will put additional pressure on world agriculture in the coming decades. The year provided more evidence that greenhouse gas emissions are rising and that climate change is already affecting agricultural productivity.

The encouraging progress made at the annual climate conventions in 2010 in Cancun and in 2011 in Durban helped address the disappointment created by the failure of the 2009 Copenhagen negotiations to result in binding commitments and gave a greater place to agriculture in global climate change negotiations. A key result was the creation of the Durban Platform for Enhanced Action. This platform, which includes all the Kyoto Protocol signatories plus the United States, is a mechanism for forging a treaty by 2015, whose goal is to bring both developed and developing countries together under a legally binding agreement by 2020.

Outside of formal negotiations, countries and regions are proceeding with their own efforts to adapt to and mitigate climate change, even in the face of a difficult macroeconomic climate. China, India, and Kenya, for instance, have all undertaken significant agricultural adaptation and mitigation activities. The progress made at the national and subnational levels should not overshadow the principle of common but different responsibilities, enshrined in the United Nations Framework Convention on Climate Change text. Rather, these national and subnational activities could be the basis of a binding multilateral agreement to pursue low-emission development strategies.

BIOFUELS

Biofuel policy changes in 2011 were dominated by the European Union, the United States, and Brazil. In the United States, the Biofuels Market Expansion Act of 2011 came into law, and debate centered on whether the Volumetric Ethanol Excise Tax Credit—a tax credit for blending ethanol into gasoline—should be repealed. Research suggests that this tax credit, combined with the ethanol blending mandate, results in both welfare and efficiency losses.[4] In addition, the Roundtable for Sustainable Biofuels was launched as a mechanism for certifying biofuel producers who adhere to standards of low environmental impact and fair labor practices. This certification could facilitate their compliance with European Union regulations and provide a "green label" that could earn them a price premium as the market further develops.

The environmental impacts of biofuel production were an important topic of investigation in the European Union during 2011. A central question concerns biofuel production and indirect land use change—that is, whether the growing use of land for biofuel crops ultimately leads to conversion of natural land to cropland, diminishing the extent to which biofuel production cuts carbon emissions. As of December 2011, the European Commission had not released its report on biofuel impacts, but once the research provides more conclusive impact findings and policy options, the region should be able to move forward with adjusting its Renewable Energy Directive.

FOOD POLICY ACTIONS & EVENTS IN 2011

WORLD ECONOMIC FORUM ON AGRICULTURE
A "New Vision for Agriculture" is presented at the World Economic Forum in Switzerland, promoting market-based solutions to accelerate sustainable agricultural growth.
January 28

CHINA NO.1 DOCUMENT
China's No. 1 Document focuses for the eighth consecutive year on water conservation and water infrastructure, due to the previous year's droughts and floods.
January 29

IFPRI NUTRITION/HEALTH CONFERENCE
More than 1,000 people attend the IFPRI-organized conference, "Leveraging Agriculture for Improving Nutrition and Health," in New Delhi, India.
February 10–12

AFRICA/INDIA FORUM SUMMIT
At the second Africa–India Forum Summit in Addis Abba, Ethiopia, "Enhancing Partnership, Shared Vision," leaders release a framework to reinforce cooperation between African countries and India.
May 24–25

G20 AGRICULTURE MINISTERS MEET
The first-ever meeting of the G20 agriculture ministers, in Paris, yields a proposal to tackle food price volatility and strengthen food security.
June 22–23

JAN FEB MAR APR MAY JUN

FOOD PRICE INDEX PEAKS

238

FEBRUARY 2011
highest peak in FPI history

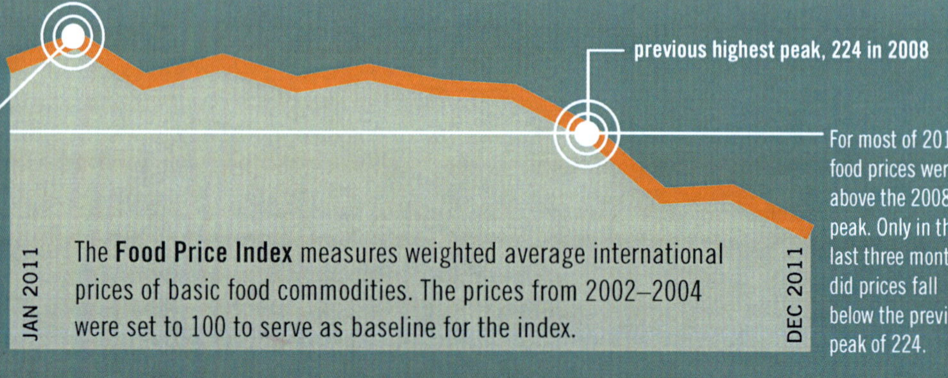

The **Food Price Index** measures weighted average international prices of basic food commodities. The prices from 2002–2004 were set to 100 to serve as baseline for the index.

For most of 2011 food prices were above the 2008 peak. Only in the last three months did prices fall below the previous peak of 224.

Brazil, China, and India have also substantially developed and revised their biofuel policies in ways that could have a large impact on food security both within their own borders and outside of them.

Finally, the 2011 disaster at Japan's Fukushima Daiichi nuclear plant revived debate on the potential drawbacks of nuclear power, and a number of countries are reducing their reliance on nuclear energy or phasing it out entirely. This debate may cause countries to shift to bioenergy, leading to further increases in global food prices.

THE FOOD AND AGRICULTURE NEXUS

In an increasingly interlinked global environment, policymakers have begun to more overtly recognize the links between agriculture and nutrition, health, water, and energy.

The agriculture, nutrition, and health nexus came to prominence in early 2011 with an international conference "Leveraging Agriculture for Improving Nutrition and Health" in New Delhi, organized by IFPRI and its 2020 Vision Initiative. This conference inspired and supported a range of new initiatives, including the launch

RUSSIA LIFTS EXPORT BAN ON GRAIN
Russia removes grain export bans put in place the previous year after wildfires destroyed a significant amount of the annual harvest.
July 1

UN DECLARES SOMALIA FAMINE
The United Nations announces that the drought in the Horn of Africa has led to outright famine in areas of Somalia.
July 20

UN FOCUSES ON NONCOMMUNICABLE DISEASES
The first-ever United Nations General Assembly on the prevention and control of noncommunicable diseases declares the need for a whole-government approach that includes the agricultural sector.
September 19–20

UN ON LAND-DEGRADATION
The United Nations General Assembly calls for building a land-degradation-neutral world, a target reflecting the green economy theme of the Rio+20 UN Conference on Sustainable Development.
September 19–20

ASEAN RICE RESERVE
ASEAN (Association of Southeast Nations) Plus Three ministers endorse the establishment of a rice emergency reserve scheme.
October 7

FOOD/NUTRITION SECURITY IN AFRICA
Africa Food and Nutrition Security Day takes place for the second time and examines "Investing in Intra-Africa Trade for Food and Nutrition Security."
October 31

BONN 2011 CONFERENCE LOOKS AT FOOD SECURITY
The German government hosts the Bonn2011 Conference on water, energy, and food security links in preparation for the Rio +20 UN Conference on Sustainable Development.
November 16–18

UN: CLIMATE CHANGE AGREEMENT
At the United Nations Climate Change Conference in Durban, South Africa, the attendees decide to adopt a universal legal agreement on climate change before 2015.
November 28–December 9

NATIONAL FOOD SECURITY BILL IN INDIA
The Indian government introduces the National Food Security Bill in parliament, shifting to a rights-based approach to food security.
December 22

JL AUG SEP OCT NOV DEC

HOW MANY WERE HUNGRY? HORN OF AFRICA FOOD CRISIS

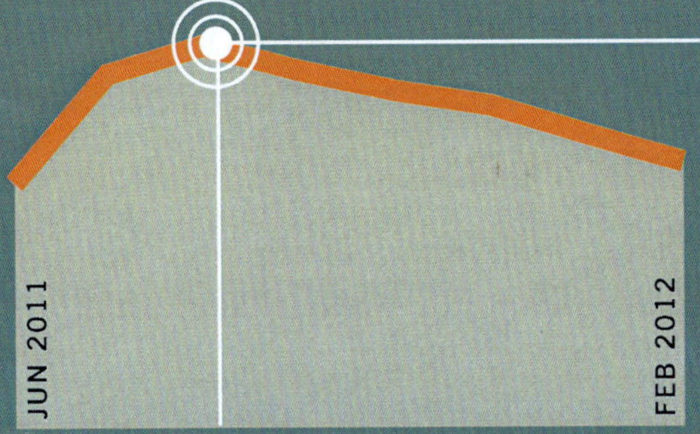

SEPTEMBER 2011
Over 13.3 million people in the Horn of Africa were affected by one of the worst droughts in 60 years.

11 million number of people targeted to receive food aid at the height of the crisis

10 months time between the first alerts about a looming crisis and the peak of the famine

SOURCE: Food and Agriculture Organization of the United Nations

of a major research program called "Agriculture for Improved Nutrition and Health" by the CGIAR. Several development agencies—USAID, with its Feed the Future Initiative, and the United Kingdom Department for International Development—also began to design or redesign their programs to better tap the links among agriculture, nutrition, and health. During 2011, 24 countries with high rates of undernutrition joined the Scaling Up Nutrition initiative, a movement bringing together governments, civil society, the private sector, research institutions, and the United Nations to support countries in their efforts to develop nutrition- sensitive national plans. More than 100 organizations also endorsed the movement. In Sub-Saharan Africa, efforts to integrate nutrition and health into agriculture development strategies were made on the continental, regional, and country level in the form of workshops, conferences, and action plans. These efforts included an agreement between the New Partnership for Africa's Development and the Global Alliance for Improved Nutrition to develop a five-year joint program to fully integrate nutrition security into the Comprehensive Africa Agriculture Development Program.

The links among food, water, and energy also gained attention in late 2011 with the conference "The Water, Energy, and Food Security Nexus" in Bonn, Germany. The Food and Agriculture Organization of the United Nations (FAO) launched a new addition to its State of the World report series with a report called *The State of the World's Land and Water Resources,* examining the availability of cultivable land, the state of land degradation, and institutions for managing scarce land and water.[5]

Despite progress, more can be done to maximize the opportunities presented by the links among agriculture and other sectors. One barrier to collaboration between agriculture and other development fields is a lack of common metrics for measuring the impact of agricultural interventions on other development outcomes such as health, nutrition, and natural resources. And more research is needed to identify viable opportunities for strengthening linkages across sectors and achieving win–win outcomes.

LAND

A rising world population, growing demand for food, fiber, and biofuels, and recent spikes in global food prices have placed increased pressure on land, resulting in more land degradation and increasing land prices, particularly in Sub-Saharan Africa, East Asia, and parts of Latin America.

Several major land policy developments transpired in 2011. The United Nations General Assembly convened a high-level meeting to address desertification, land degradation, and drought, with government representatives highlighting not only the threat posed by land degradation to social, economic, and environmental sustainability, but also the need for future investment in sustainable land management. Several initiatives—specifically, the FAO's Global Soil Partnership as well as the Economics of Land Degradation initiative undertaken by Germany, the European Commission, and the United Nations Convention to Combat Desertification—were launched as mechanisms for strengthening sustainable land management through knowledge building and sharing. New evidence presented at these events by IFPRI researchers shows that policymakers should pay attention to land degradation not just in dry areas, but also on many high-quality irrigated lands. More should be done to assure the availability of fertilizers in areas where additional fertilizer use is needed and appropriate to improve soil fertility.

One dimension of land management policies that particularly occupied public discourse in 2011 was the issue of foreign land acquisitions—often described as "land grabbing"—especially in Sub-Saharan Africa. Such acquisitions have the potential to inject much-needed investment into agriculture in developing countries, but they can also harm the food security and livelihoods of the local poor. Large-scale land deals may also have negative impacts on gender equity if they erode women's customary land rights.[6] Reports on the issue in 2011 by the FAO, the World Bank, and the International Fund for Agricultural

Development all highlighted the need for governments to ensure responsible investment in agriculture and to strengthen land administration systems that respect the rights, livelihoods, and resources of all citizens.[7]

NEW PLAYERS

New "players"—such as the private sector, emerging economies, and philanthropic organizations—are increasingly reshaping the structure and nature of the global food policy landscape. Not only are these new players a largely untapped source of financial support to food security efforts in developing countries, but they also offer a wealth of knowledge and expertise, providing new opportunities to address the increasing complexity and challenges facing the global food system.

In 2011 these new players became more entrenched in global food policymaking processes. For example, the G20 is quickly claiming a growing role, next to the G8, as a principal forum for managing global economic problems. The action plan of the G20 agriculture ministers also emphasized the importance of strengthening the engagement of nonstate actors, especially the private sector, in global food security efforts. Emerging economies such as Brazil, China, and India have increased their engagement, especially in terms of forging South–South cooperation. In 2011, for example, the FAO and China made three-party agreements with Liberia and Senegal to provide Chinese technical assistance to food security initiatives and projects. One noteworthy development has been the initiation of cooperation agreements between the Bill & Melinda Gates Foundation and emerging economies such as Brazil and China in support of agricultural and health innovations in the developing world.

Other 2011 initiatives demonstrate the private sector's increasing involvement in global food security efforts. The World Economic Forum released a "Roadmap for Stakeholders" as part of its New Vision for Agriculture Initiative. This initiative—a collaboration among the World Economic Forum's partner companies—promotes market-based strategies for sustainable agricultural development. In parallel, the Forum's partner CEOs contributed to

New "players"—such as the private sector, emerging economies, and philanthropic organizations—are increasingly reshaping the structure and nature of the global food policy landscape.

the development of policy positions on food price volatility and food insecurity that fed directly into the 2011 deliberations of the G20 agriculture ministers. Public–private partnerships have been launched to promote sustainable agricultural growth, reduce hunger, and improve nutrition. For instance, PepsiCo has signed several agreements with international organizations to support increased agricultural production (especially among smallholders) alongside long-term nutritional and economic security efforts in countries such as China, Ethiopia, and Mexico. Similarly, private philanthropic and civil society organizations have continued to be major supporters of agricultural development, nutrition, poverty alleviation, and natural resource management.

Still, the opportunities presented by these new players have not been fully harnessed. For example, the private sector's presence in many global food security platforms is essentially limited to multinational corporations, and there is no real platform for engaging smaller companies. And until recently, the traditional aid donor community—represented by the Organisation for Economic Co-operation and Development's Development Assistance Committee—has not involved new players.

REGIONAL DEVELOPMENTS

Some regional developments shaped food security and agriculture, as well as development more broadly, over the course of 2011.

In parts of North Africa and the Middle East, long-standing factors—ranging from youth unemployment to growing income disparities and high risk of food insecurity—led to the Arab Spring, mainly in Egypt, Libya, and Tunisia, but also in Bahrain, Syria, and Yemen. Addressing the challenges that gave rise to the Arab Spring will require more inclusive development strategies. To improve household food security, governments in the region will need to adopt policies that stimulate inclusive growth, such as employment generation for the young and poor, as well as expanded and well-targeted safety nets.

African countries made significant progress in implementing the Comprehensive Africa Agriculture Development Programme (CAADP) in 2011. This program is the African Union's continent-wide framework to boost agricultural productivity and food security. Six countries signed compacts committing them to achieving an agricultural sector growth rate of 6 percent a year and to raising funding for the sector to at least 10 percent of the national budget—bringing the total number of signatory countries to 29. About 20 of these countries have developed national investment plans, and 6 have received funding totaling US$270 billion from the Global Agriculture and Food Security Program.

In India, Parliament introduced the National Food Security Bill, which would provide rice, wheat, and coarse grains at low prices to more than half of India's 1.2 billion people, making it the world's largest antihunger program. China announced plans to boost agricultural productivity through increased public investments in water conservation and irrigation. Its water conservation investments will total about US$630 billion over the next 10 years.

In Central America and the Caribbean, high and volatile prices and natural disasters raised concerns about "a hungrier" region. In October 2011, the ministers of agriculture of the Americas approved a declaration emphasizing the importance of increasing investment in agriculture to reduce hunger and poverty and help improve social stability in the hemisphere.

In Europe and the United States, continued policy support to biofuel production, farm subsidies, a hostile attitude toward agricultural biotechnology (mainly in Europe), and trade protections have negatively affected the agriculture sector in developing countries.

OUTLOOK FOR 2012 AND OPPORTUNITIES FOR ACTION

Overall, 2011 and the years immediately preceding it have revealed serious weaknesses facing the global food system—lack of ability to respond to volatile food prices, extreme weather, and inadequate response to food emergencies were among the most visible. But chronic, long-term problems such as food and nutrition insecurity also point to areas where the food system can do better. We also face uncertainties. It is not yet clear whether the global economic slowdown will worsen or be reversed. Addressing all of these issues in a resource-scarce world will require keeping agriculture and food security issues high on the global agenda in 2012 and beyond.

Without preventive action, several hot spots could erupt in food crisis in 2012. Early warning systems are once again pointing to the risks posed by drought in Africa—this time in the Sahel region, including Burkina Faso, Mali, Niger, and Senegal. The experience in the Horn of Africa was a tragic reminder of the need to move quickly and aggressively to head off humanitarian crises. Uncertainty also surrounds North Korea, long a recipient of food aid, which is undergoing a leadership transition.

Participants in the major international events of 2012 need to keep the spotlight on food policy issues. The G8 summit in the United States in May and the G20 Summit in Los Cabos, Mexico, in June could reinforce those groups' earlier emphasis on global food security and ensure that previous financial commitments are honored. It is important that discussions and decisions at the Rio+20 conference on the green economy and sustainable development not neglect the poor, who need better access to food, jobs, and natural resources, as well as a secure social protection system.

More broadly, food policy decisionmakers will face a number of challenges in 2012 and beyond. The long-term problems of chronic food and nutrition insecurity persist, although they are

← Looking Back
Looking Forward →

ENCOURAGING EVENTS IN 2011

- Agriculture, nutrition, and health climbed up on the national and global agendas, and the nexus of agriculture, food, land, water, and energy has received more attention (see Chapter 6).
- The world's major political leaders made food policy a high priority, with the G20 agreement on an Action Plan on Food Price Volatility and Agriculture.
- At the World Economic Forum, the world's business and society leaders gave agriculture a boost when they initiated their New Vision for Agriculture.
- Encouraging progress was made at the climate change conference in Durban, acknowledging the role agriculture can play in the mitigation of and adaptation to climate change (see Chapter 4).
- China's focus on agricultural policy bore fruit as total grain production exceeded 570 million tons, a new record (see Chapter 9).
- India's Parliament introduced a National Food Security Bill to provide affordable grains to more than half of its 1.2 billion people (see Chapter 9).
- New initiatives like Feed the Future, the Global Agriculture and Food Security Program, and South-South cooperations boosted agriculture investments.
- Promoting mother and child nutrition gained momentum as it became widely accepted that the nutrition in the 1,000 days between conception and a child's second birthday are of crucial importance for the child's future.

NOT WHAT WE HOPED FOR IN 2011

- High and extremely volatile food prices in the first half of the year threatened the food security of millions of people (see Chapter 2).
- Biofuel policies in the United States and the European Union have not been changed to take into account their impact on land-use change and food price volatility (see Chapter 5).
- The Doha Round of trade negotiations was still not finalized, so countries continued to maintain domestic policies that undermine the trading prospects of developing countries and the sustainability of the global food system.
- Setting a clear international standard or "code of conduct" for large-scale foreign investment in land has received too little attention.
- African countries are not meeting their target of allocating at least 10 percent of national budgetary resources to agriculture.
- The international community responded slowly and too late to the disaster that was unfolding in the Horn of Africa (see Chapter 3).
- Hunger still persists globally: nearly one billion people go hungry every day. The 2011 Global Hunger Index indicates that more than two dozen countries have "alarming" or "extremely alarming" hunger levels.

WHAT TO WATCH FOR IN 2012

- How are governments responding to financial crises and how does this affect their development assistance, especially in the fields of agriculture and nutrition security?
- How much progress is being made on the various initiatives taken in 2011, like the G20 Action Plan or the G8's repeated commitment to improve food security?
- What impact are noncommercial transactions in futures markets and the increasing trading volume of index funds having on high and volatile prices of agricultural commodities? (See Chapter 2.)
- To what extent is agriculture being integrated in environmental and sustainability discussions, including EarthSummit 2012 or the ongoing climate change debate?
- What are the new leaders of the World Bank, the Food and Agriculture Organization of the United Nations, and the World Food Programme doing to promote nutrition security and agriculture?
- Are the lessons learned during the crisis in the Horn of Africa being applied to increase effectiveness and impact when addressing the emerging crises in the Sahel and North Korea?
- How is the balance of power shifting in agricultural research, technology, production, and trade, with emerging economies pushing the agricultural agenda? (See Chapter 8.)
- Which countries are making the most progress toward achieving the first Millennium Development Goal, and why?

sometimes overshadowed by more dramatic events and acute crises. We will soon reach the 2015 target date of the Millennium Development Goals, almost certainly without having met the goal of halving hunger globally. South Asia and Sub-Saharan Africa, in particular, still show alarming levels of food and nutrition insecurity, despite the progress achieved in recent years. In addition, more work will be needed to reach an effective international agreement on climate change.

We must find new ways to exploit the links between agriculture and other sectors, including health, nutrition, water, and energy. Paying attention to gender equity will help make investments and interventions in these areas more effective. Because agriculture is at the nexus of all of these areas, we need to leverage it for broad development outcomes. At the same time, it will be important to set up a global system to measure, track, and monitor the impacts among agriculture, food and nutrition security, energy, and natural resources. In addition, to allocate resources more effectively, we should begin to base the prices of natural resources and food on their full value to society, including their social and environmental costs, such as impacts on climate change and health. All of these actions require skills and knowledge at the national and local level, so capacity building can help improve outcomes.

These events and challenges will play out in different ways in each country. National and local policies are where global forces translate into on-the-ground impact, so good governance and effective leadership and implementation can make a big difference. Some countries would benefit greatly from a stronger emphasis on building the capacity—that is, the skills and knowledge—of policymakers and program implementers at all levels.

This outlook points to some high-priority areas for action in 2012. First, the G20 should take further steps to rein in food price volatility by, for example, doing more to reduce the competition between biofuel and food production and to discourage trade restrictions that exacerbate price swings. Second, the international community should consolidate global and regional agricultural growth strategies and create or strengthen the institutions and capacities needed to make these strategies work. In particular, this year's G8 summit should work to ensure that the industrial countries meet their financial commitment in support of a country-led development process for achieving food security in developing countries. Third, participants in the Rio+20 meeting should integrate economic, social, and environmental sustainability efforts and commit to concrete action to meet the long-term challenges of development, including poor nutrition, degraded soils, and scarce water. Finally, a broad intersectoral coalition should work together to address issues related to nutrition, food, and health. ■

What Influenced Food Policy in 2011?

Most spectacular in 2011 was the turn of events on world wheat markets from price spike to near collapse: In the spring the media expected a second world food crisis, possibly worse than 2007–08. Until July, and particularly head of the meeting of G20 agricultural ministers, speculators and index funds were being accused more than ever of causing hunger. But then wheat prices dropped, and attention to speculation waned, hopefully making room for policy attention to larger, more long-term issues, such as rural finance.

—Michiel A. Keyzer, Director, Centre for World Food Studies, VU University, Amsterdam

Amid drought in the Horn of Africa, floods in South East Asia, and rain shortfalls in the Sahel, 2011 has clearly shown the devastating impact of climate-related shocks on food security. These crises have focused policy attention on the urgent need to build the resilience of smallholder agriculture and poor rural people's livelihoods. Going forward, and in light of the UN Climate Change Conference in Durban, resilience is likely to remain a critical component of food security policies, initiatives, and development efforts at all levels.

—Kanayo F. Nwanze, President, International Fund for Agricultural Development, Rome

The Arab Spring posed the biggest challenge to food policy in 2011—and showed why it matters. Arab countries are squeezed on all sides by high imported food prices, spiraling costs of food subsidies, and the dual burdens of malnutrition and obesity, which will rise with population growth. The region is also the most vulnerable to global warming, water scarcity, and export bans. Without good policy and research, feeding the Arab world will grow ever more challenging.

—John Parker, Globalization Editor, Economist, London

When food prices rose in 2008, hasty responses like banning food exports helped drive 100 million people into poverty—the first increase in decades. When food prices rose again in 2011, the world avoided poor policy responses and invested instead in long-term food security. During the world's worst drought in 60 years, this approach was validated by Kenya and Ethiopia's ability to avoid famine, thanks in part to President Obama's Feed the Future initiative and its emphasis on building resilience through agricultural development.

—Rajiv Shah, Administrator, United States Agency for International Development, Washington, DC

In 2011 two events were important: one was the eighth consecutive year of bumper harvest of Chinese grains at a record of 571 million tons, which surely contributes to a more stable world grain market; and the other was the G20 Agriculture Ministers Summit in Paris. A new era of international cooperation on global food security is approaching and emerging countries such as Brazil, China, India, and Indonesia will play increasingly important roles.

—Jiayang Li, President, Chinese Academy of Agricultural Sciences, Beijing

The developing world was again hit by food price and supply volatility in 2011. In contrast to 2008, the demand for effective actions to advance food and nutritional security was front and center. The Committee on World Food Security explicitly stated that agricultural policies and public investment should prioritize nutrition and sustainable small-scale food production and increase the resilience of local and traditional food systems and biodiversity, a goal we are fully committed to implementing.

—Kathy Spahn, President and Chief Executive Officer, Helen Keller International, New York

The Democratic Republic of the Congo, which contains 75 percent of the world's second largest rainforest, wants to be a leader in reducing emissions from forests. Financing is expected to run in the billions of dollars, which demonstrates the government's increased commitment to agriculture. Speculation in agricultural commodities was also high on the agenda in 2011. There is little evidence that speculators systematically drive food prices, but they do affect price volatility. However, limiting speculative trading might do more harm than good. The G20 decided to create more transparency and asked the UN's Food and Agriculture Organization to monitor trading more closely.

—Eric Tollens, Professor Emeritus, Katholieke Universiteit, Leuven, Belgium

For the first time the G20 placed a high priority on agriculture. Price volatility and food security were priorities of the French presidency. Interest in these issues continues into 2012 under the Mexican presidency and is likely to generate significant investments in agriculture, thus addressing declining productivity.

—Justin Yifu Lin, Senior Vice President and Chief Economist, World Bank, Washington, DC

Persistent high food prices, among other things, triggered the formation of land markets, leading to excessive commercial pressure on land in a context of ill-defined property rights. A new landscape of energy policy emerged—shale gas, bioenergy, and partial exits from atomic energy in Germany and Japan. It comes with indirect linkages to agriculture (in the form of opportunity costs) and raises challenges to address climate change. Food policy was also significantly advanced by the G20 debate and proposals to increase agriculture aid, commodity trading improvements, and the related US and European follow-up that will accommodate more transparency and less speculation.

—Joachim von Braun, Director, Department for Economic and Technological Change, Center for Development Research, Bonn, Germany

Climate-smart agriculture increases productivity, strengthens farmers' resilience, and reduces agriculture's contribution to climate change by reducing greenhouse gas emissions and increasing carbon storage on farmland. Growing global recognition of climate-smart agriculture and its potential to offer triple wins for food security, adaptation and mitigation was one of the major success stories of 2011, and has real potential to influence national food policy.

—Rachel Kyte, Vice President of Sustainable Development, World Bank, Washington, DC

The G20 process, with the creation of the Agricultural Market Information System and general recognition of the importance of better information significantly influenced food policy in 2011. So did the growing acceptance of the UN Food and Agriculture Organization's findings (in the 2011 State of Food and Agriculture report) that promoting gender equality and equity would bring the number of hungry down by 150 million. Also FAO's launch of a new agricultural paradigm, "Save and Grow," which is designed to increase global food production sustainably.

—José Graziano da Silva, Director General, Food and Agriculture Organization of the United Nations, Rome

The increasing momentum of the Scaling Up Nutrition movement was evident in 2011. The movement supported country-led efforts to improve nutrition through cooperative partners working across sectors toward a common goal. Scaling Up Nutrition promotes both direct nutrition interventions and nutrition-sensitive strategies such as improving agricultural practices to increase availability of nutrient-rich crops. The 2011 international conference "Leveraging Agriculture for Improving Nutrition and Health," coordinated by the 2020 Vision Initiative of IFPRI, sparked the interests of global counterparts and served as a timely complement to the Scaling Up Nutrition collective effort.

—Emorn Wasantwisut, Senior Advisor, Institute of Nutrition, Mahidol University, Salaya, Thailand

I am pleased with last year's extraordinary commitment by world leaders to improve human nutrition, which has stimulated the emergence of a country-led movement to "Scale Up Nutrition." I am particularly impressed with the way this has engaged a broad range of stakeholders and is encouraging nutrition-sensitive agricultural, industrial, health, education, employment, social welfare, and economic policies. I welcome the focus on improving the coverage of specific actions to improve nutrition from conception to a child's second birthday and on political accountability for equitable improvement in nutrition within the context of policies for food, health, and social security.

—David Nabarro, Special Representative of the United Nations Secretary General on Food Security and Nutrition, New York

For the eighth consecutive year, China's total grain production increased, reaching 571 million tons last year and exceeding 550 metric tons for the first time in half a century. This helped China fight domestic consumer-price inflation and stabilize world food prices. Also, a study group headed by Yuan Longping, China's father of hybrid rice, announced that the yield of hybrid rice per Mu exceeds 900 kilogram in one of its trial sites. This would contribute greatly to Chinese and world food security.

—Keming Qian, Director General, Department of Development and Planning, Ministry of Agriculture, Beijing

In 2011 Oxfam launched its most ambitious campaign: GROW. Food prices, flattening yields, climate change, unfair trade, failing markets, inequality between men and women and land grabs are all connected and contributing to a global food system that is dominated by a few powerful governments and companies, while failing the majority of people. GROW will push policy and practice changes from the global to local levels to grow more food more fairly and sustainably.

—Jeremy Hobbs, Executive Director, Oxfam International, Oxford, England

The destabilizing effects and uncertainties created by the recent price hikes of major staple foods and the food crises and famine in the Horn of Africa, have raised food security concerns to a higher political level, receiving more attention and priority consideration than in the past in the agendas of decisionmakers in governments. This is an important step forward, since food security is a highly political issue that requires political solutions, rather than a humanitarian issue that needs technical solutions as it was often seen in the past.

—Carlos Pérez del Castillo, Chair, CGIAR Consortium Board, Montpellier, France

The importance of an integrated approach to food security that IFPRI has helped prioritize is vital in today's world. The year 2011 and the famine in the Horn of Africa reinforced the role of social safety net programs in providing a broad package of support for the most vulnerable—from specialized nutrition products to protect the minds and bodies of young children, to investments in sustainable land management to help communities' build resiliency to drought.

—Josette Sheeran, Executive Director, World Food Programme, Rome

In Canada, the most important food policy event was influenced by ideology rather than market or resource policy shifts: the government's decision to abolish the Canadian Wheat Board which for decades has sold all Western Canadian wheat. This will open up new market opportunities for the international wheat majors. On water issues, there were interesting indications that the Indian national government is looking for the political and financial space to assume a larger role, for example, by including major irrigation canal investments in its next five-year plan.

—Margaret Catley-Carlson, Chair, Crop Diversity Trust, Rome, and Patron, Global Water Partnership, Stockholm

In our 2011 World Disasters Report, the IFRC addressed one of the most persistent critical issues facing our word today: hunger. As an Ethiopian, I saw first-hand my country's terrible famine and I know what it means for people to starve. Globally, an estimated 925 million people do not have enough to eat, and as the population grows between now and 2050, global food supplies will come under even greater pressure. Governments must acknowledge the right to food and implement comprehensive, community-centered hunger prevention programs now and increase equitable and sustainable investments in food security.

—Bekele Geleta, Secretary General, International Federation of Red Cross and Red Crescent Societies, Geneva

The G20 focused on food security and price volatility and led to international research initiatives to secure an adequate level of production. The Wheat Initiative was decided to promote highly productive wheat systems adapted to climate change. The GEO-GLAM project aims to monitor cultivated areas in order to predict harvests, as better anticipation prevents the formation of "bubbles" in agricultural markets. In 2011, G20 decisions represented a major step forward in coordinating efforts to improve World Food Security.

—Marion Guillou, Chief Executive Officer, French National Institute for Agricultural Research, Paris

Chapter 2

FOOD PRICES

Riding the Rollercoaster

Maximo Torero, IFPRI

THE WORLD FACES A NEW FOOD ECONOMY that likely involves both higher and more volatile food prices, and evidence of both phenomena was on view in 2011. After the food price crisis of 2007–08, food prices started rising again in June 2010, with international prices of maize and wheat roughly doubling by May 2011. The peak came in February 2011, in a spike that was even more pronounced than that of 2008, according to the food price index of the Food and Agriculture Organization of the United Nations. When prices of specific commodities are adjusted for inflation, though, the 2011 price spike did not reach the levels of 2008 (Figure 1).

Although the food price spikes of 2008 and 2011 did not reach the heights of the 1970s, price volatility—the amplitude of price movements over a particular period of time—has been at its highest level in the past 50 years. This volatility has affected wheat and maize prices in particular. For hard wheat, for example, there were an average of 27 days of excessive price volatility a year between January 2001 and December 2006 (according to a measure of price volatility recently developed at IFPRI[1]). From January 2007 to December 2011, the average number of days of excessive volatility more than doubled to 76 a year (Figure 2).[2]

High and volatile food prices are two different phenomena with distinct implications for consumers and producers. High food prices may harm poorer

FIGURE 1 Inflation-adjusted prices of agricultural commodities and oil, 1990–2011

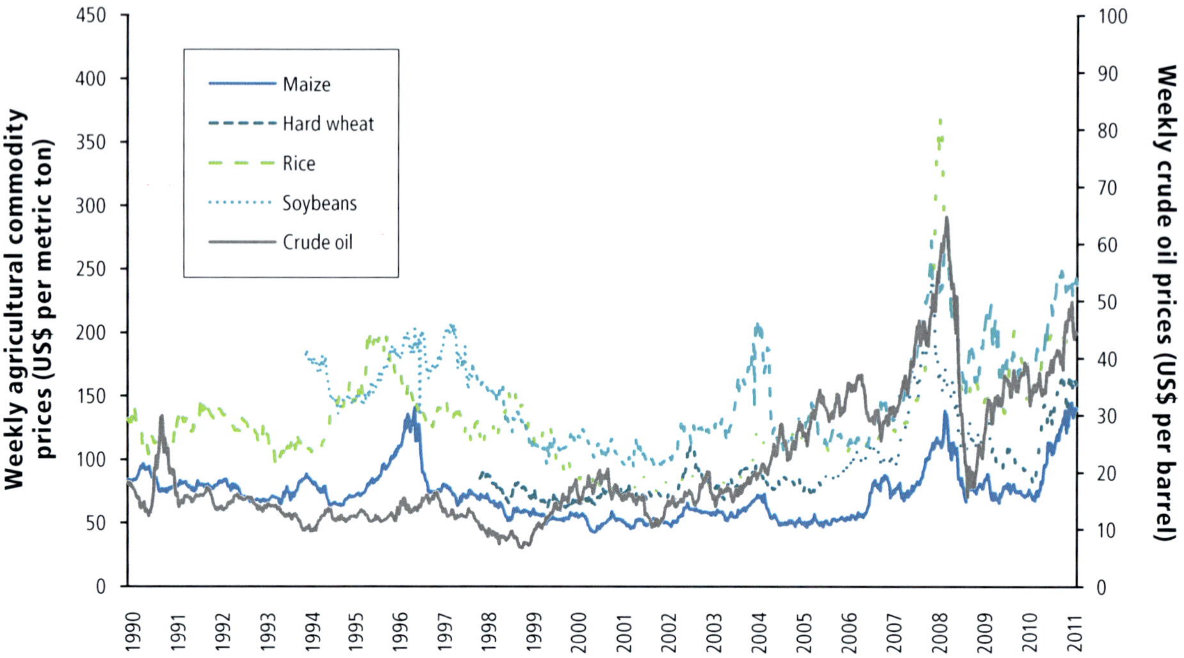

Source: Food and Agriculture Organization of the United Nations, FAOSTAT database, http://faostat.fao.org/; International Grains Council, *World Grain Statistics 2009* (London, 2009); US Energy Information Administration, World Crude Oil Prices (online data) (Washington, DC, 2011), www.eia.gov. **Note:** Prices are adjusted for inflation using a consumer price index base year of 1982–84 (that is, 1982–84 = 100). Maize is US No. 2 Yellow, wheat is US No. 2 Hard Red Winter, rice is White Thai A1 Super, soybeans are US No. 1 Yellow, and crude oil is the spot price for West Texas Intermediate at Cushing, Oklahoma.

consumers because they need to spend more money on their food purchases and therefore may have to cut back on the quantity or the quality of the food they buy or economize on other needed goods and services. For food producers, higher food prices could raise their incomes—but only if they are net sellers of food, if increased global prices feed through to their local markets, and if the price developments on global markets do not also increase their production costs. For many producers, particularly smallholders, some of these conditions were not met in the food price crisis of 2011.

Apart from these effects of high food prices, price volatility also has significant effects on food producers and consumers. Greater price volatility can lead to greater potential losses for producers because it implies price changes that are larger and faster than what producers can adjust to. Uncertainty about prices makes it more difficult for farmers to make sound decisions about how and what to produce. For example, which crops should they produce? Should they invest in expensive fertilizers and pesticides? Should they pay for high-quality seeds? Without a good idea of how much they will earn from their products, farmers may become more pessimistic in their long-term planning and dampen their investments in areas that could improve their productivity. (The positive relationship between price volatility and producers' expected losses can be modeled in a simple profit maximization model assuming producers are price takers. Still, it is important to mention that there is no uniform empirical evidence of the behavioral response of producers to volatility.) By reducing supply, such a response could lead to higher prices, which in turn would hurt consumers.

It is important to remember that in rural areas the line between food consumers and producers is blurry. Many households both consume and produce agricultural commodities. Therefore, if prices become more volatile and these households reduce their spending on seeds, fertilizer, and other inputs, this may affect the amount of food available for their own consumption. And even if the households

FIGURE 2 Excessive food price variability for hard wheat

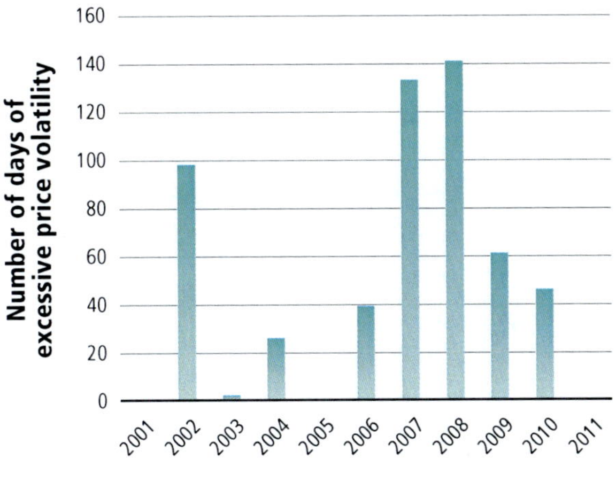

Source: C. Martins-Filho, F. Yao, and M. Torero, "High Order Conditional Quantile Estimation Based on Nonparametric Models of Regression," www.foodsecurityportal.org/sites/default/files/martins-filho_torero_yao_2011_0.pdf. Accessed April 3, 2012.

are net sellers of food, producing less and having less to sell will reduce their household income and thus still affect their consumption decisions.

Finally, increased price volatility over time can also generate larger profits for investors, drawing new players into the market for agricultural commodities. Increased price volatility may thus lead to increased—and potentially speculative—trading that in turn can exacerbate price swings further.

DRIVERS BEHIND RECENT FOOD PRICE VOLATILITY

Among the key factors playing a role in creating price volatility are increasing biofuel production, the medium- and long-term effects of climate change, and higher levels of trading in commodity futures markets. Export restrictions in important food-producing countries also contributed to price increases and market jitters in 2010 and 2011.

Biofuel policies. With oil prices at significantly high levels during 2011, and with the European Union and the United States subsidizing and setting mandates for biofuel production, farmers have shifted their cultivation toward biofuel crops, most of which are also used as food or feed, such as maize, sugar, and oilseeds (Figure 3). To comply with biofuel mandates, farmers have ramped up production of such crops, increasing the demand for land, water, and nutrients—and therefore the production costs of other food crops. Furthermore, the production of biofuel crops strengthens the links between the highly volatile energy markets and food markets, thereby increasing the volatility of food prices. With more countries, such as India and Peru, enacting biofuel mandates, food price volatility is likely to increase even further. Flexible biofuel mandates that will not contribute to food price volatility could represent alternative mechanisms to reduce the potentially negative impact of biofuel policies.[3]

Extreme weather and climate change. Extreme weather events helped raise food prices and fuel price volatility in 2007–08 and 2010–11,[4] and climate scenarios predict more variable weather events in the future.[5] More intense and frequent natural disasters (such as droughts and floods) resulting from climate change could trigger significant yield losses and subsequent price increases and higher volatility. Indeed, IFPRI simulations show that climate change is likely to push prices up, regardless of whether population (and thus demand for food) grows faster or slower. In contrast to the 20th century, when inflation-adjusted prices of staple grains declined, in the first half of the 21st century, these prices are likely to rise.[6]

Commodity futures trading. One signal of higher price volatility has been the significant

FIGURE 3 Share of US maize crop used to produce ethanol, 1995–2010

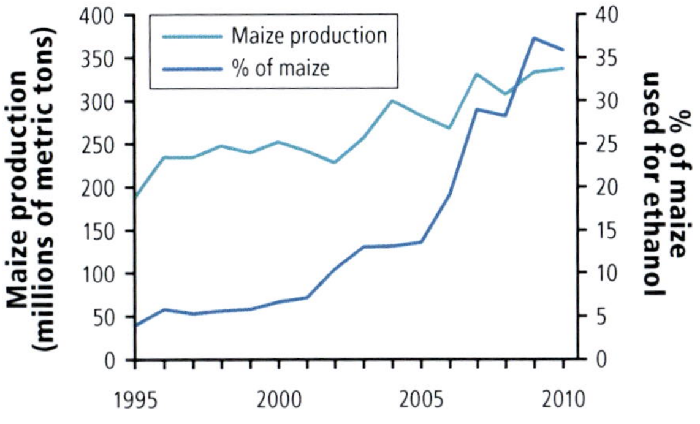

Source: Data from Earth Policy Institute, Data Center: Climate, Energy, and Transportation (2011), www.earth-policy.org/data_center/C23.

increase in the volume of agricultural commodity futures traded in the Chicago Board of Trade, a leading agricultural futures exchange. (Futures are contracts between a buyer and a seller that specify a current price for a commodity to be delivered on a certain date in the future. These contracts can themselves be traded by investors who do not physically own the commodity or plan to take delivery of it.) From 2005 to 2006, the average monthly volume of futures trading for wheat and maize grew by more than 60 percent. In 2007, traded volumes again rose significantly for wheat, maize, rice, and soybeans. In fact, the average monthly volume of trading in soybean futures was 40 percent larger than in 2006 (Figure 4). Futures trading continued to increase during 2010–11 for all commodities. Between March 2006 and December 2011, the volume of commodity index funds trading increased (in terms of the number of transactions of 5,000 bushels) by 157 percent, 200 percent, and 169 percent for maize, soybeans, and soft wheat at the Chicago Board of Trade and by 124 percent for hard wheat at the Kansas City Board of Trade. Investors have increased their trading of food commodity futures, but only 2 percent of these futures contracts have resulted in the delivery of real goods. For maize, for example, the volume of futures traded on exchanges worldwide is more than three times greater than the global production of maize.

Changes in futures prices have been shown to lead to changes in day-to-day, or "spot," prices. This pattern of increasing commodity futures trading and higher prices for commodity futures can create a vicious circle that exacerbates the volatility of spot prices for food commodities to excessive levels.[7]

Other factors. Today's agricultural markets have three characteristics that make the price responses to these challenges more extreme. First, export markets for the main staple commodities—rice, maize, wheat, and soybeans—are either highly concentrated in a few countries or very "thin" (that is, only a small share of production is traded) (Figure 5). Given these high levels of concentration, the world's capacity to cope with shocks is limited. Any incidence of poor weather or other production shocks in these countries will immediately affect global prices and price volatility. Similarly, any policy changes—such as trade bans, customs taxes, or other restrictions on exports—in any of the top exporters will significantly affect the levels and volatility of food prices (see Figure 6). Research suggests that such policies explained almost 40 percent of the increase in the world market price for rice during the 2007–08 food price crisis.[8]

Second, the world's stocks of cereals are now at historically low levels (Figure 7). This situation leaves the world vulnerable to food price spikes and threatens the proper functioning of markets. The world's cereals stocks, measured as a ratio of stocks

FIGURE 4 Monthly volume of futures trading, 2002–11

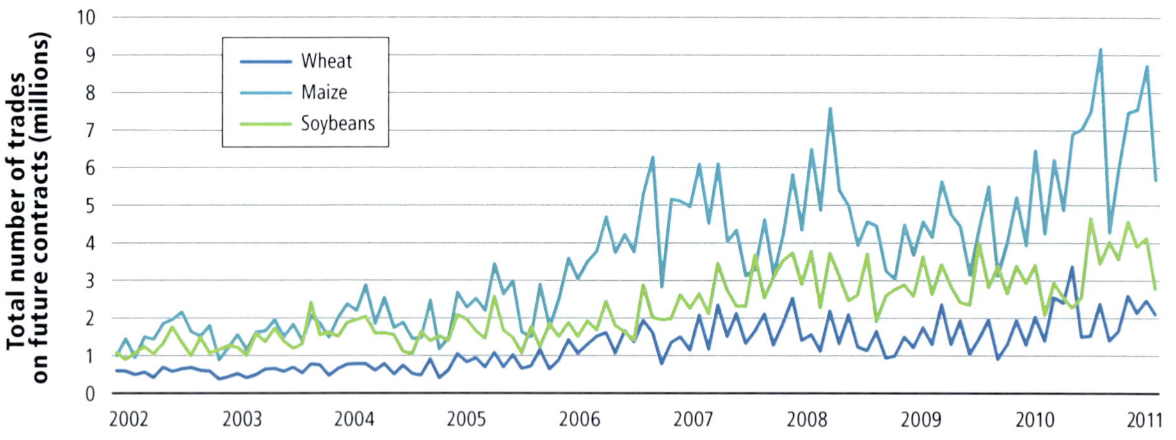

Source: Chicago Board of Trade.

FIGURE 5 **Major exporters' shares of global maize, wheat, and rice exports, 2008**

MAIZE 84%	WHEAT 63%	RICE 95% (paddy)	RICE 85% (milled)
United States - 53.0%	United States - 22.9%	United States - 90.4%	Thailand - 36.4%
Argentina - 15.1%	France - 12.4%	Paraguay - 1.4%	Vietnam - 19.9%
Brazil - 6.3%	Canada - 12.0%	France - 1.2%	Pakistan - 10.9%
France - 6.0%	Russian Federation - 8.9%	China - 1.1%	India - 10.4%
India - 3.5%	Argentina - 6.7%	Brazil - 0.9%	United States - 7.2%

Source: Food and Agriculture Organization of the United Nations, FAOSTAT database, http://faostat.fao.org/. **Note:** Paddy rice is mostly exported by the United States, and milled rice is exported by other countries.

to cereals use, were similarly very low when wheat prices spiked in the 1970s, 1995–96, 2007–08, and 2010–11. This indicates that for the market to function effectively, the food system must hold a minimum level of grain stocks to be able to respond to unexpected shocks (such as bad weather) and allow for the transport, marketing, and processing of grains.[9] Given the current low levels, sometimes only a small dip in grain stocks leads to problems. In 2007–08 grain stocks were only about 60 million tons less than in 2004–05, representing a decline of just 2.7 percent of global production. But when prices rose sharply in 2007–08, this difference in grain stocks was enough to contribute to serious price increases, especially for commodities whose production is concentrated in just a few countries, such as rice.[10]

FIGURE 6 **Effects of trade policy reactions for selected countries on world wheat prices**

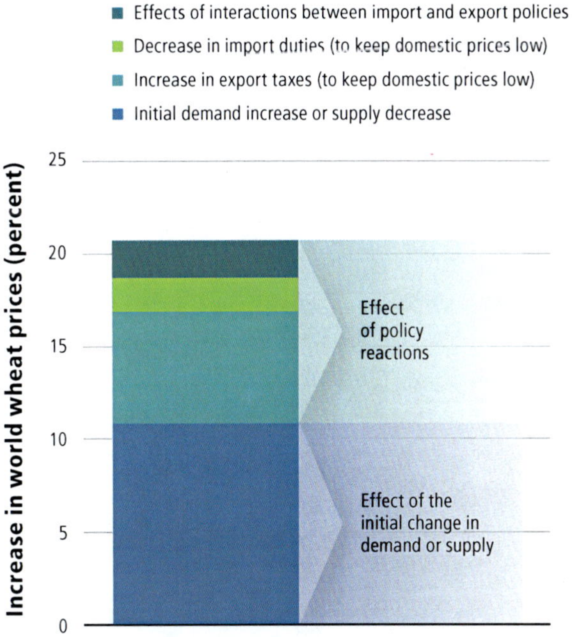

Source: A. Bouet and D. Laborde, *Economics of Export Taxation in a Context of Food Crisis: A Theoretical and CGE Approach Contribution*, IFPRI Discussion Paper 994 (Washington, DC: International Food Policy Research Institute, 2010).

FIGURE 7 **Ratio of cereals stocks to use, 1996/97–2011/12**

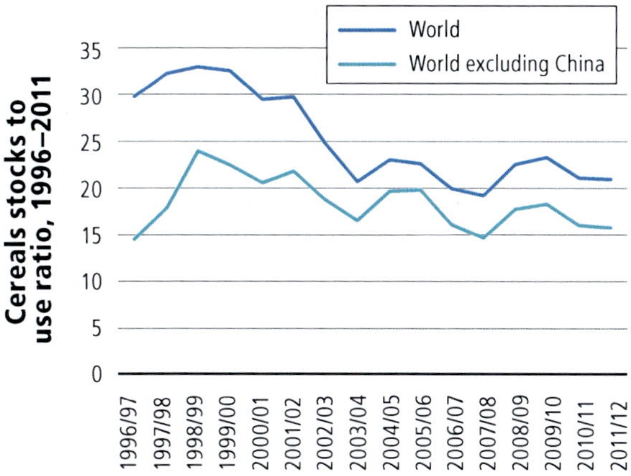

Source: Food and Agriculture Organization of the United Nations, *Food Outlook* (Rome, various years). **Note:** World excluding China is shown because China is an outlier in terms of reserves, and there are several questions about the quality of its reports.

Third, appropriate and timely information on food production, stock levels, and price forecasting is sorely lacking. When information deficits lead to overreactions by policymakers, the result can be soaring prices.

ACTIONS AND PROPOSALS

In the wake of the two recent food price crises, some actions have been taken and many proposals have been put forward to prevent such events from occurring again. These can be grouped by the objectives they try to achieve: (1) better information and more research, (2) easier trade in agricultural commodities, (3) larger food reserves and better-managed grain stocks, (4) more active use of financial instruments to influence agricultural commodity markets, and (5) stricter regulation of these markets. Scholars and policymakers are debating the merits, feasibility, and likely effectiveness of many aspects of these proposals.

Better information and more research. Recognizing the need for better information, the Group of 20 (G20) countries agreed in June 2011 to launch the Agricultural Market Information System (AMIS). The AMIS is designed to encourage major players in global agrifood markets to cooperate more and to share data and information. If it is properly linked to existing early warning systems at global and national levels, it could substantially improve countries' ability to make sound decisions on food security and help reduce price volatility. To make the AMIS effective, countries and regions need to develop transparent and publicly accessible systems for monitoring food security and collecting data so that they can provide appropriate information on food production balances and reserves. In addition, the system will require the full participation of private agrifood companies, which hold much of the world's stocks of grain. So far, private companies are merely urged to participate in the AMIS on a voluntary basis, and without their participation the information will be incomplete and the system will have limited impact.

Two other key proposals have been made to improve information and coordination in ways that would increase market confidence and relieve temporary disruptions in supply. The first, from Brian Wright and Alex Evans, is for an international food agency,[11] following the example of the International Energy Agency. This food agency would report on stock levels and develop protocols for the global response to shortages to help prevent market panic. Two criticisms have been directed at this proposal. First, generating better information on stocks will likely involve considerable effort and international coordination—and therefore cost. Currently, information is lacking not only on the public stocks held by key producer countries such as China and India, but also on the stocks held by private enterprises, which consider them commercial secrets. Second, with this information asymmetry, it is not clear how the proposed agency would identify the threshold of stocks at which international collaboration would be required or how countries would agree on an emergency response.

The second proposal, from Carlos Martins-Filho, Maximo Torero, and Feng Yao, consists of an early-warning mechanism for identifying abnormally high price volatility in the futures prices of staple food crops on a daily basis.[12] This information could help reduce the potential asymmetry of information between buyers and sellers and thereby help dampen price volatility. There is one main caveat for this model: it currently operates only for commodities traded on the futures market, but it could be extended to spot markets if better price information existed.

Easier trade in agricultural commodities. In the 2007–08 and 2010–11 food price crises, many countries responded by cutting exports or boosting imports in ways that worsened price increases. Some proposals therefore aim to facilitate trade to reduce risks in grain trading when supplies are low and to avoid disruptions in global grain markets. One proposal is for a food import financing facility that would help poor countries afford food imports at times of high prices, as well as an international grain clearinghouse arrangement to ensure the availability of staple food imports.[13] This clearinghouse would guarantee contracts for grain deliveries, reducing the risk that exporters would renege on contracts when supplies are tight. In a different

BOX 1

Trade Restrictions Amplify Food Price Spikes

Kym Anderson, University of Adelaide

In 2011 export bans continued to hurt poor people. Governments often raise import barriers during turbulent times to mitigate immediate domestic concerns such as unemployment, but during the recent global financial and food price crises some countries raised export barriers.[1] Such government action aims to make exporting food more difficult and expensive, thereby protecting domestic consumers from the effects of an international food price spike. Examples in 2011 were bans on grain exports in Tanzania, Ethiopia, and Russia—all of which were lifted before the end of the year. Such responses exacerbate the price spike (by restricting supplies in the international market) and affect the international welfare transfer associated with that spike in terms of trade (which defines how much a country needs to export in exchange for a given import volume).[2]

Much less appreciated is the fact that governments of food-importing countries are equally concerned for their consumers. As a result, many of them lowered their food import restrictions, and some even switched to food-import subsidies. This further exacerbated the international price spike—which meant it weakened the initial attempt by food-exporting countries to shield their consumers.

New evidence on the extent of the change in domestic relative to international prices in food exporting and importing countries reveals that

- historically, only around half the movement in international food prices is transmitted to domestic markets within the first year;
- both grain-exporting and grain-importing countries react to food price spikes with a similar speed and on a similar scale when restricting trade;
- the changes in restrictions on global grain trade during 2006–08 are responsible for estimated increases in the international prices of rice, maize, and wheat of around two-fifths, one-fifth, and one-tenth, respectively;
- domestic prices of wheat would have risen less on average across all countries if trade restrictions had not been changed; and
- altered trade restrictions caused rice price increases in both high-income and developing countries to be only one-quarter to one-third less than what they otherwise would have been.

The policy conclusion is this: in our globalizing world, attempts to insulate domestic consumers from international food price spikes are mostly futile. Those actions hurt all food-importing countries by increasing the price of their imports. Stronger World Trade Organization disciplines on both exports and imports are clearly needed to limit how much damage such beggar-thy-neighbor government responses can do in the global marketplace when food prices spike.

approach, other observers propose preventing export bans to avoid any disruption of supplies.[14]

Larger food reserves and better-managed grain stocks. Proposals have been put forth for physical reserves, including emergency reserves,[15] international coordinated grain reserves,[16] regional reserves, and country-level reserves.

An emergency reserve is a modest stock of about 300,000–500,000 metric tons of basic grains—about 5 percent of current global food aid flows—which would be supplied by the main grain-producing countries and funded by a group of more than a dozen countries. This reserve, to be used exclusively for emergency response and humanitarian assistance, could be managed by the World Food Programme. In 2011, in response to this proposal, the G20 proposed studying the feasibility of a global humanitarian emergency reserve through a pilot program in West Africa under the leadership of the Economic Community of West African States and with the support of the World Food Programme.

Global or regional reserves will require a trigger mechanism that determines when to release stocks to calm markets in times of stress, and it is essential that such a mechanism be transparent. The proposed early warning system for price volatility, mentioned earlier, could be a solution.

Finally, a physical reserve, whether regional or global, will not resolve the problem of links among the financial, energy, and food commodity markets. This is a key problem that could be extremely relevant if excessive speculation is indeed a cause of extreme price spikes.

More active use of financial instruments. Two major proposals are linked to the use of financial instruments: (1) virtual reserves[17] and (2) a toolbox of market-based risk management tools.

A virtual reserve would involve intervening in futures markets based on price volatility data from the early warning mechanism already described or, in extreme cases, a decision by a technical committee. This intervention would consist of executing a number of progressive short sales (that is, selling a firm promise to deliver the commodity at a later date at a specified price) until futures prices and spot prices decline to specified acceptable levels. This increase in short sales would reduce spot prices and should lower extreme price volatility by cutting the probability of abnormal returns. Most of the time, futures contracts would be settled through offsetting purchases or sales—in other words, the whole operation would be virtual. Only rarely would it be necessary to obtain the necessary grain supply to comply with futures contract delivery requirements. A virtual reserve has several advantages compared with a physical reserve: it is just a signaling mechanism; it does not put more stress on commodity markets; it does not incur the significant storage and opportunity costs of a physical reserve; it resolves the problem of the interlinkages between the financial and the commodity markets; and given that it is only a signal, it should have only a minimal effect on markets.

The toolbox, proposed in the 2011 meeting of the G20 ministers of agriculture, would include mechanisms such as physical or financial commodity price hedges, insurance, and guarantee instruments, as well as countercyclical lending, which could help vulnerable countries mitigate the risks associated with excessive food price volatility. Two initiatives are being implemented. The first, under the management of the International Finance Corporation, involves a new Agriculture Price Risk Management tool that will allow producers and consumers to hedge against downside or upside price risks on a pilot basis. The second is a World Bank proposal to facilitate governments' access to risk management markets. It entails helping to structure and execute financial and physical commodity risk hedging and to build capacity related to the legal, regulatory, and technical requirements associated with using these tools. Both of these initiatives will need to be evaluated to ensure their effectiveness, viability, and sustainability.

Stricter regulation. Since late 2005 problems have plagued the futures and cash markets for maize, soybeans, and wheat. The main problem is lack of convergence between cash and futures prices. To address this issue, the US Commodity Futures Trading Commission, other agencies in the US government, and the European Commission, along with the futures industry, have moved forward with setting seasonal storage rates, imposing limits on the number of delivery certificates an entity can hold for noncommercial purposes, and putting out an additional issue of the Commitments of Traders report to increase transparency. For example, in October 2011 the US Commodity Futures Trading Commission approved caps on speculation in food, energy, and metals, restricting the size of positions to 25 percent of deliverable supply. If the structural changes put in place do not significantly improve the price convergence between futures and cash prices, then a cash-settled contract must be seriously considered.

CONCLUSIONS

The global food price crises of 2007–08 and 2010–11 led to economic difficulties for the poor, contributed to political turmoil in many countries, and in the long run could undermine confidence in global food markets, thereby hampering these markets' performance in balancing fundamental changes in supply, demand, and production costs. More important, food price crises can result in unreasonable or unwanted price fluctuations that can harm the poor, especially by compromising their nutrition security. One consequence is long-term, irreversible nutritional damage, especially among children. Therefore these recent food market episodes highlight the need to reform the architecture of international financial and agricultural markets to address the problem of price spikes and protect the most needy and vulnerable.

In response to the food price crises, a mix of policy actions have been taken. Many countries have tried to build up costly national reserves, and others have focused on increasing self-sufficiency. Still others have lost confidence in the reliability of food trade in global markets, which has led some countries to acquire farmland overseas to ensure national food security. In addition, some countries are pressing for more regulation of commodity exchanges—however, whether this would prevent extreme price spikes or instead distort markets even further is questioned. All of these policy actions threaten to move food and agriculture markets further away from efficient arrangements. A more promising step may be regionally coordinated reserves, as recently planned by the Association of Southeast Asian Nations. This global problem needs international institutional responses.

The analysis here points to three clear messages. First, we need to respond to the structural problems faced by the agricultural sector—that is, the concentration of global exports of staple grains among just a few exporters, the low levels of global grain stocks, and the lack of appropriate information. Second, it is crucial to evaluate the effects of policies designed to promote biofuels and investments in derivative markets, such as commodity futures, as well as the limited actions taken to cope with the risks of climate change, such as weather insurance. Finally, we will need to carefully monitor many current actions being taken to reduce the frequency of price spikes and excessive volatility to assess how cost-effectively they cope with the new developments in global food markets. ■

BOX 2

Rethinking the Role of Food Reserves

Peter Timmer, Professor Emeritus, Harvard University

In 2011, the world again saw proof that large food reserves dampen the volatility of food prices, as large rice stocks in India kept global rice prices from following wheat and maize prices to record high levels. Although they are costly to maintain, larger food reserves provide supplies in times of crisis. More importantly, in vulnerable countries, reserves build confidence that trade remains the most efficient mechanism for stabilizing domestic food economies. Low levels of foodgrain reserves, on the other hand, make commodity markets nervous and subject to sudden demand and supply shocks—and even to speculative activities. Therefore, if less volatile food prices are desired, two questions remain: How large should grain reserves be? And who should own them?

Private markets have a clear and coherent answer to the first question, but only if governments stay out of the business of holding grain stocks. Long-standing models show that optimal storage levels exist when price expectations match the expected returns from holding grain in storage. Unfortunately, with regards to ownership, foodgrain stocks held in private hands are usually insufficient to provide a politically acceptable level of food security, especially in large countries. This typically results in governments stepping in to stabilize domestic food prices, using one of two basic methods: (1) imposing restrictions on food trade, which tends to increase price volatility in world markets, or (2) enabling public ownership of food reserves, which can be expensive.

The evidence supporting the need for large grain reserves clearly exists, but collective action at the global level is not likely. Helping countries build up their own domestic reserves, however, is possible. Larger reserves will help stabilize the global food economy and thus allow trade to play a larger (and less disruptive) role. If the international development community, in partnership with governments of large countries, wants a more stable global food economy, we need to change the long-run incentives for stockholding behavior and use increased stocks to build confidence in the role of the international market for foodgrains. Because holding larger stocks will turn out to be very expensive, a scenario can be imagined where the larger stocks are built gradually and steadily create renewed confidence in the world grain market as prices become more stable. Stocks will then be reduced (gradually) as the reality of the fiscal burden sinks in. What should remain is the renewed trust in trade and how it can help even large countries sustain their food security.

UILDIN
CAL
RHINO
RHINO
UILDING A

DISASTERS

Déjà Vu in the Horn of Africa

Derek Headey, IFPRI

THE YEAR 2011 WILL BE REMEMBERED FOR some of the most severe "natural" disasters on record. There were major natural disasters in both developed and developing countries (see "Food Security & Food Safety" map on the following pages): powerful earthquakes in Japan, New Zealand, and Turkey; major floods in Pakistan (see Box 3), Southeast Asia, and Australia; and significant droughts in the Horn of Africa and parts of the Sahel. Within this list, there is substantial diversity in terms of the severity of the shocks, in whether they were slow-moving or sudden-onset disasters, in whether the shocks were effectively one-time events or a more regular feature of the landscape, and in whether the societies affected by the disaster were relatively resilient or relatively vulnerable.

In lowland areas of the Horn of Africa, droughts and floods are frequent events, although the scale of the 2011 food emergency was somewhat unusual. The drought began with failed rains in late 2010 and mid-2011. In some parts of the Horn of Africa—particularly parts of Somalia—the drought was the worst in 60 years. Moreover, at the peak of the drought—around August 2011—more than 13 million people were in need of food assistance. The United Nations Children's Fund reported that more than 320,000 children were suffering from severe malnutrition in Djibouti, Ethiopia, Kenya, and

FOOD SECURITY & FOOD SAFETY
WHERE DISASTERS STRUCK IN 2011

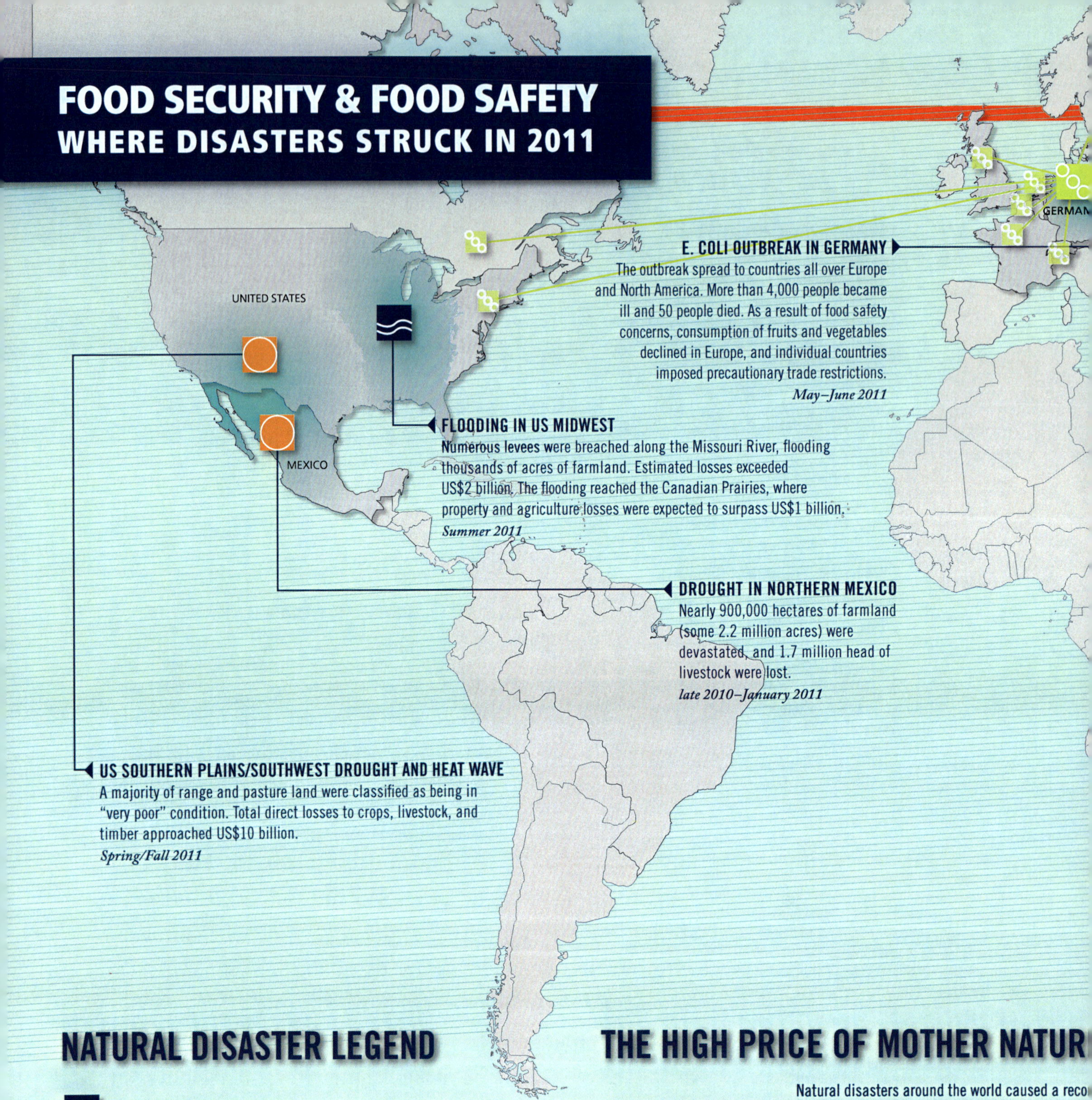

E. COLI OUTBREAK IN GERMANY
The outbreak spread to countries all over Europe and North America. More than 4,000 people became ill and 50 people died. As a result of food safety concerns, consumption of fruits and vegetables declined in Europe, and individual countries imposed precautionary trade restrictions.
May–June 2011

FLOODING IN US MIDWEST
Numerous levees were breached along the Missouri River, flooding thousands of acres of farmland. Estimated losses exceeded US$2 billion. The flooding reached the Canadian Prairies, where property and agriculture losses were expected to surpass US$1 billion.
Summer 2011

DROUGHT IN NORTHERN MEXICO
Nearly 900,000 hectares of farmland (some 2.2 million acres) were devastated, and 1.7 million head of livestock were lost.
late 2010–January 2011

US SOUTHERN PLAINS/SOUTHWEST DROUGHT AND HEAT WAVE
A majority of range and pasture land were classified as being in "very poor" condition. Total direct losses to crops, livestock, and timber approached US$10 billion.
Spring/Fall 2011

NATURAL DISASTER LEGEND

- FLOODING
- CYCLONE
- DROUGHT
- EARTHQUAKE
- TSUNAMI
- FOOD-BORNE DISEASE
- FAMINE/FOOD SCARCITY

THE HIGH PRICE OF MOTHER NATUR

Natural disasters around the world caused a reco
US$380 billion in economic losses in 2011. That's
more than twice the tally for 2010, and about
US$115 billion more than the previous record.

$380 BILLION

$152 BILLION

$210 BILLION
more than half of total losses in 2011

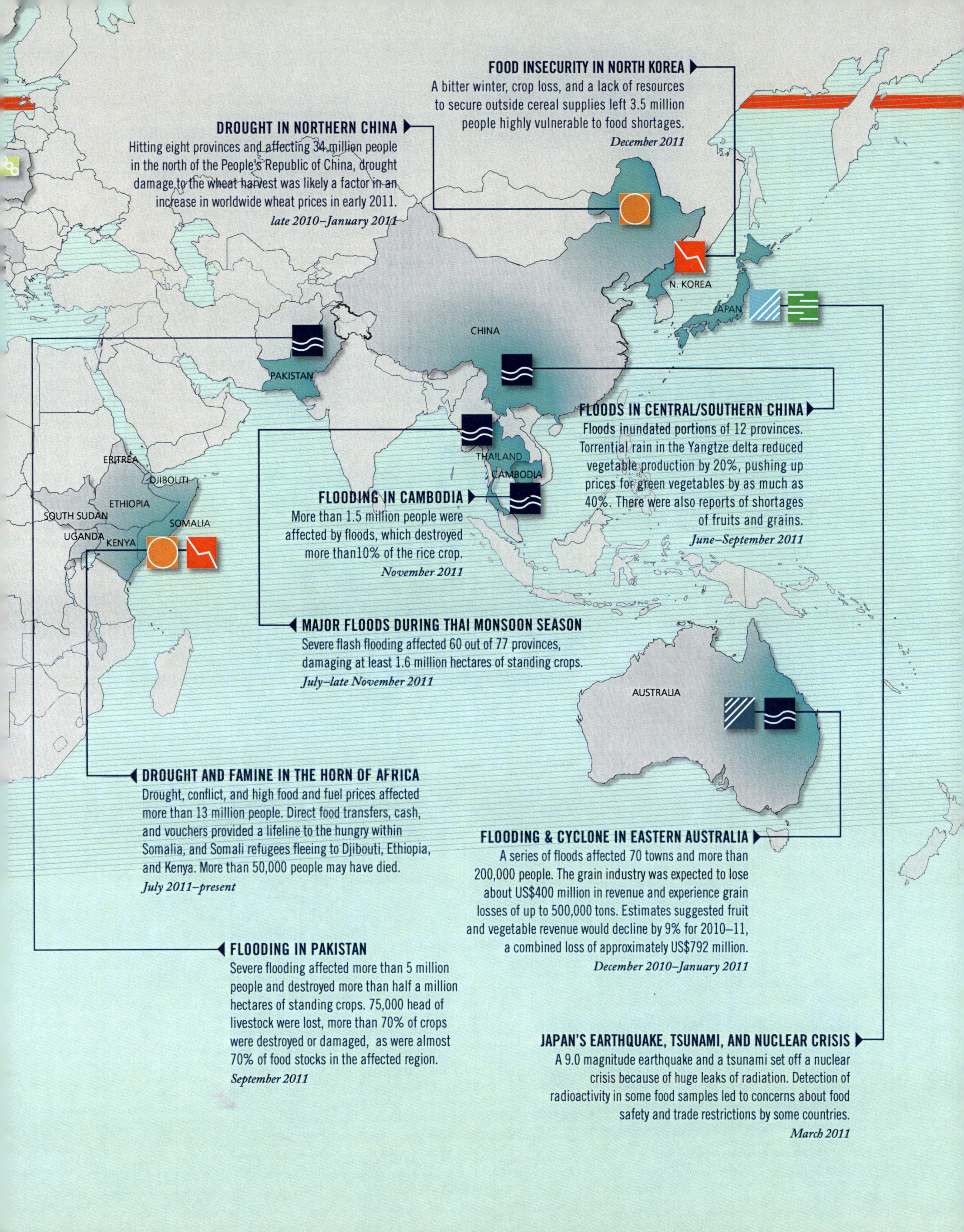

FOOD INSECURITY IN NORTH KOREA

A bitter winter, crop loss, and a lack of resources to secure outside cereal supplies left 3.5 million people highly vulnerable to food shortages.

December 2011

DROUGHT IN NORTHERN CHINA

Hitting eight provinces and affecting 34 million people in the north of the People's Republic of China, drought damage to the wheat harvest was likely a factor in an increase in worldwide wheat prices in early 2011.

late 2010–January 2011

FLOODS IN CENTRAL/SOUTHERN CHINA

Floods inundated portions of 12 provinces. Torrential rain in the Yangtze delta reduced vegetable production by 20%, pushing up prices for green vegetables by as much as 40%. There were also reports of shortages of fruits and grains.

June–September 2011

FLOODING IN CAMBODIA

More than 1.5 million people were affected by floods, which destroyed more than10% of the rice crop.

November 2011

MAJOR FLOODS DURING THAI MONSOON SEASON

Severe flash flooding affected 60 out of 77 provinces, damaging at least 1.6 million hectares of standing crops.

July–late November 2011

DROUGHT AND FAMINE IN THE HORN OF AFRICA

Drought, conflict, and high food and fuel prices affected more than 13 million people. Direct food transfers, cash, and vouchers provided a lifeline to the hungry within Somalia, and Somali refugees fleeing to Djibouti, Ethiopia, and Kenya. More than 50,000 people may have died.

July 2011–present

FLOODING & CYCLONE IN EASTERN AUSTRALIA

A series of floods affected 70 towns and more than 200,000 people. The grain industry was expected to lose about US$400 million in revenue and experience grain losses of up to 500,000 tons. Estimates suggested fruit and vegetable revenue would decline by 9% for 2010–11, a combined loss of approximately US$792 million.

December 2010–January 2011

FLOODING IN PAKISTAN

Severe flooding affected more than 5 million people and destroyed more than half a million hectares of standing crops. 75,000 head of livestock were lost, more than 70% of crops were destroyed or damaged, as were almost 70% of food stocks in the affected region.

September 2011

JAPAN'S EARTHQUAKE, TSUNAMI, AND NUCLEAR CRISIS

A 9.0 magnitude earthquake and a tsunami set off a nuclear crisis because of huge leaks of radiation. Detection of radioactivity in some food samples led to concerns about food safety and trade restrictions by some countries.

March 2011

SOURCES: Food Security and Food Safety: Earth Policy Institute; European Centre for Disease Prevention and Control; *The Guardian*; United States Department of Commerce National Oceanic and Atmospheric Administration; Oxfam; PricewaterhouseCoopers; National Committee for Disaster Management (Cambodia); Food and Agriculture Organization of the United Nations; United Nations World Health Organization; United Nations World Food Programme; World Bank. **The High Price of Mother Nature:** Munich Re.

Somalia. The situation in Somalia was particularly grave: 4 million people—more than half of the country's population—were in crisis (Figure 1). Of these, 750,000 were officially declared as experiencing famine. Since mid-2011, thousands are known to have died, especially infants and small children.

Given the severity of this drought, and the frequency of humanitarian emergencies in the region, several troubling questions arise. Why is the region seemingly more vulnerable now than in the past, especially after decades of humanitarian and development assistance? And what steps need to be taken to improve development and relief efforts to render the peoples of the Horn more resilient to the next drought?

WHY IS THE HORN OF AFRICA SO VULNERABLE?

Unlike some of the other disasters of 2011, the crisis in the Horn of Africa is not a one-time event. Since the Great African Famine of 1982–84, vulnerability (Figure 2) and aid dependency appear to have

BOX 3

After the Floods: Pakistan and Food Security

Paul Dorosh and Sohail J. Malik, IFPRI

Unusually heavy monsoon rains contributed to severe flooding in parts of Pakistan in 2010 and 2011. From July to August 2010, flood waters covered 50,000 square kilometers and affected more than 18 million people (about one-tenth of the total national population), resulting in about 2,000 flood-related deaths, loss of 500,000 livestock, and damage to or destruction of 2.2 million hectares of standing crops, 1.7 million homes, and 10,000 schools. Then, close on the heels of this disaster, the 2011 flood struck southern parts of Pakistan in August through October. Although it affected a smaller area than the 2010 flood and only about half as many people, the combined human and physical costs underscored the importance of continued improvements in disaster rehabilitation and recovery in Pakistan.[1]

Earlier experiences in Pakistan and other South Asian countries have shown that disaster recovery should incorporate livelihood strategies for affected households, including

- prioritizing social protection of the most vulnerable groups,
- raising awareness about new programs,
- ensuring the participation of key stakeholders (from a multisector base as well as the community) in the decisions made for each program,
- tailoring interventions to specific needs of vulnerable groups, and
- providing temporary work schemes.

In particular, the experience of the 1998 Bangladesh flood—where poor households had a continuing debt burden of about US$100 (equivalent to a month and half's average consumption) even fifteen months after the flooding—highlighted the importance of private-sector borrowing in the coping strategies of the poor and the need to consider substantial transfers to these households to avoid long-term adverse effects.

While Pakistan benefited from some of the experiences of the past, delays in funding and implementation—caused in part by donor reluctance in the face of a deteriorating governance and law and order situation—plagued the response to the Pakistan 2010 floods. Thankfully, domestic wheat prices in Pakistan remained stable due to a good harvest in April 2010 and abundant private and public stocks. The Pakistani authorities processed 1.5 million flood-affected households and provided almost 900,000 households with emergency shelter. About 6 million people received food assistance in monthly rations through January 2011. The Government of Pakistan also initiated a Citizen's Damage Compensation Program designed to give to each of the 1.5 million affected families a one-time payment of approximately US$230 (in the form of a debit card or "Watan Card"). An ex post evaluation of program effectiveness, including targeting of payments and other aspects, should yield additional useful lessons and insights for future disaster preparedness and relief and recovery in Pakistan.[2]

FIGURE 1 Estimated food insecurity at the height of the Horn of Africa famine

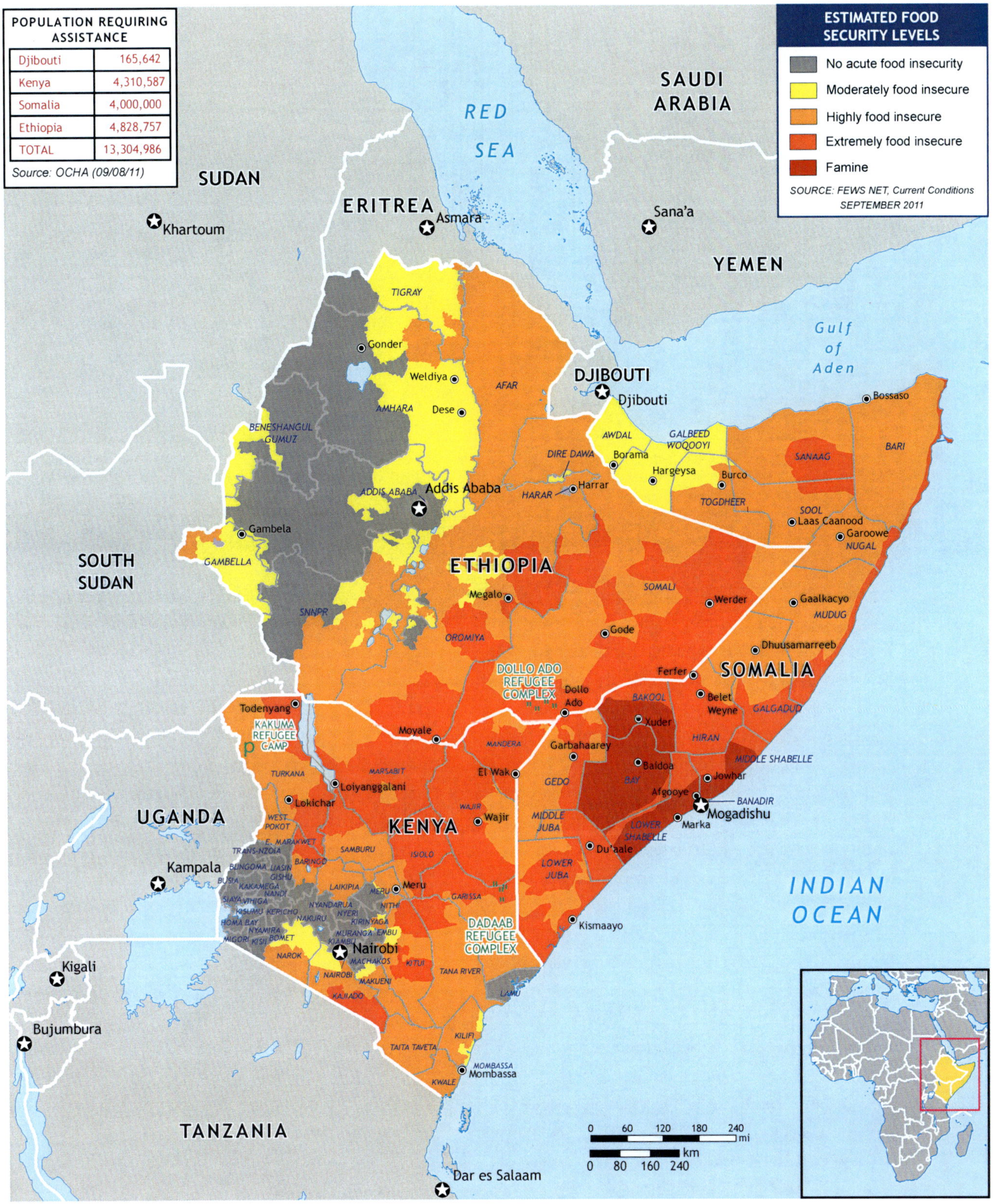

Source: US Agency for International Development, http://www.usaid.gov/our_work/humanitarian_assistance/disaster_assistance/countries/horn_of_africa/template/maps/fy2012/hoa_10072011.pdf. **Notes:** Data for this map were estimated for September 2011.

FIGURE 2 Number of people adversely affected by droughts in the Horn of Africa, 1970–2010

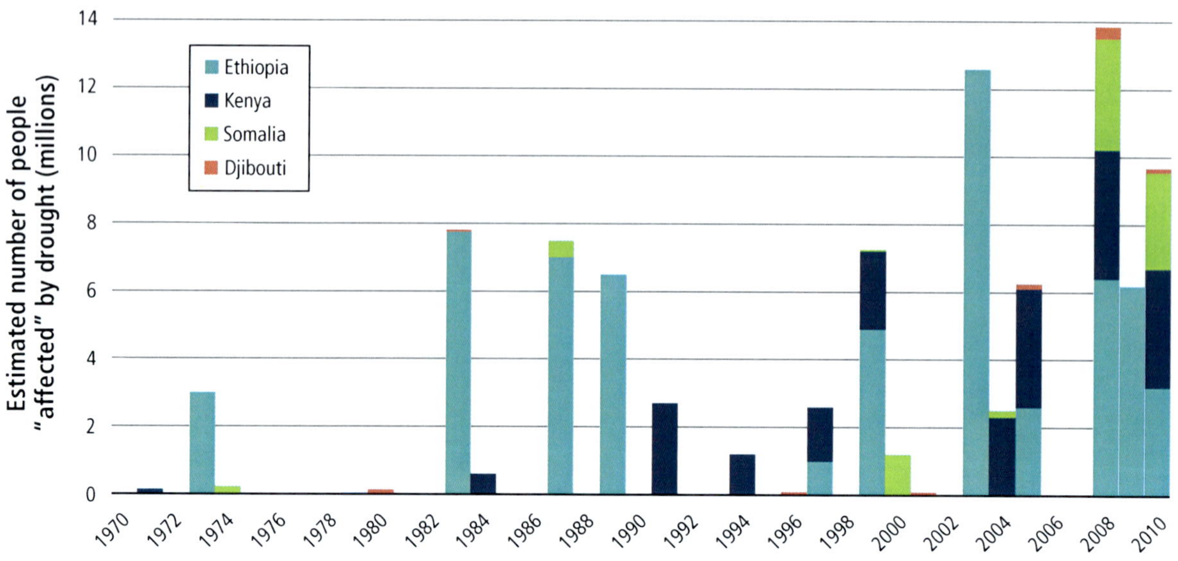

Source: EM-DAT: The OFDA/CRED International Disaster Database (Université Catholique de Louvain, Brussels, 2011), www.emdat.be. **Notes:** The estimates here are only approximate. In addition to the problems of identifying affected people, there are degrees of impact that are not recorded, as well as possible omissions in earlier periods due to less effective measurement, poor governance, and so on. Also, these data do not distinguish between pastoralist and nonpastoralist populations, particularly in Ethiopia, making it difficult to gauge the impacts of droughts in pastoralist areas specifically.

increased over time. In Kenya, droughts accompanied by food emergencies occurred in six of the past eight years. But what explains this disturbing trend? Is it because droughts and floods are more frequent, because people are more vulnerable, or both?

There is not yet evidence of widespread climate change in the Somali region of Ethiopia,[1] but rainfall in Kenya appears to have declined substantially, and some observers predict that climate change will soon increase drought frequency in the region.

Even so, most experts on the region see the apparent increase in food insecurity as a function of socioeconomic factors as well as climatic events. This thinking partly reflects previous research on famines and food insecurity, stemming from the seminal work of Nobel laureate Amartya Sen.[2] Sen hypothesized that people starve not because of aggregate food shortages, but because they cannot get access to food. But other observers argue that famine also has deeper social and political causes, such as conflict, corruption, and other forms of economic and political mismanagement.[3]

In the most recent crisis, most people agree that a major reason that southern Somalia alone was gripped by famine was the conflict in that region (Figure 1). The conflict has likely increased food insecurity through several channels. First, without an effective government, southern Somalia has not been able to develop the kinds of disaster risk management institutions and social safety-net programs found in other countries in the region. Second, Al Shabab has excluded the World Food Programme from the areas it controls, greatly inhibiting the supply of emergency goods and services (Box 4). And third, conflict has significantly constrained people's mechanisms for coping with drought, such as their ability to move their herds and to engage in trade to sell off livestock and get access to affordable food supplies. Local conflict has been shown to inhibit herd mobility in various regions of Ethiopia and Kenya.[4]

Not only can conflict amplify the effects of drought, but drought can cause conflict by exacerbating competition over scarce grazing lands and water supplies. For Somalia a recent study argues that rainfall shortages push down real livestock prices (and therefore household incomes), which in turn leads to more frequent conflict as young men

BOX 4

Humanitarian Aid: How Can We Do Better?

Steven Were Omamo, World Food Programme

Drought, conflict, and high food and fuel prices affected the lives of more than 13 million people in the Horn of Africa region—Djibouti, Ethiopia, Kenya, Somalia, and Uganda—in 2011. Working closely with governments and other partners, the World Food Programme (WFP) targeted 11 million people affected by the crisis. By December, employing a range of interventions, including direct food transfers, cash, and vouchers, WFP had reached almost 8 million people across the region, providing a critical lifeline to vulnerable Somalis within Somalia in particular, and also to Somali refugees fleeing to Djibouti, Ethiopia, and Kenya. The bulk of WFP's food assistance reached drought-affected populations in Ethiopia, Kenya, and Uganda. However, insecurity and poor infrastructure within Somalia prevented WFP from reaching all targeted populations.

Despite falling short of its goal, an important lesson for WFP emerged. Investments by national governments and other partners in improved land-use management and other resilience-enhancing measures ensured that populations that required food assistance during previous droughts did not need such support in 2011, in particular in Ethiopia, Kenya, and Uganda.

For WFP, investments in enhanced preparedness also paid off, especially decisions to use a newly created advance-purchasing facility to acquire and pre-position food in areas likely to require food assistance. Preliminary analysis by WFP, the African Union, and other partners suggests operation of regional emergency food reserves and expanded use of weather-index insurance could further improve preparedness in situations such as the one in the Horn of Africa.[1]

Looking ahead, key policy challenges facing WFP and other humanitarian actors center on how to strengthen the resiliency of communities living in drought-prone areas, using humanitarian assistance to help farmers and pastoralists adapt to changes in weather patterns. To that end, agencies must find ways to

- better integrate relief efforts into longer-term solutions that build resilience among communities in drought-prone areas, expanding scope for recovery and rehabilitation;
- protect productive assets of affected populations, with a special focus on meeting the nutritional needs of the weakest members of society by providing highly nutritious supplementary food products;
- strengthen the capacity of national governments to develop institutional arrangements and mechanisms to address crises, with an emphasis on approaches that balance short-term interventions with medium- and long-term investments that address low productivity and other causes of food and nutrition insecurity; and
- support the African Union Commission and the Intergovernmental Authority for Development in creating a process to address critical regional policy and institutional gaps, especially by more effectively linking assessment and early warning alerts with timely and effective action.

These actions can limit the negative effects of a natural or human-caused crisis, which will reduce suffering and increase the impact of humanitarian aid.

look for alternative forms of income.[5] Unbundling the exact relationships between drought, conflict, and food insecurity is difficult, but it seems likely that conflict is both a cause and consequence of food insecurity.

Conflict is an obvious socioeconomic explanation of food insecurity in the Horn, but it is by no means the only one. Many studies of the Horn—particularly outside Somalia—have focused on the declining resilience of pastoralists and ex-pastoralists.[6] Households' resilience is chiefly a function of their assets (livestock, education, land) and their coping mechanisms (mobility, income diversification). Livestock is the largest economic sector in the Horn, and for many households, it is the most important asset and an important source of income and milk and other products for their own consumption. Given the region's abundant land and variable rainfall, moving livestock from place to place has traditionally been an effective way of coping with drought.

Historically, however, there has been a long-standing debate on whether mobile livestock rearing is sustainable. Some early critics argued that the difficulty of managing common resources led to excessive herd build-up and boom-and-bust cycles, overgrazing and land degradation, and depletion of water resources.[7] In the 1990s and 2000s, however, a growing body of evidence suggested that herd build-up in post-drought years was a rational attempt to increase overall herd resilience to subsequent droughts.[8] There is also now a fairly broad consensus that pastoralism does not lead to permanent damage to rangelands.[9]

Given the Horn of Africa's abundant land and variable rainfall, moving livestock from place to place has traditionally been an effective way of coping with drought.

Yet this does not mean that the issue of the region's "carrying capacity" is no longer relevant. Even if pastoralists' herd management strategies are individually rational and ecologically sound, human and livestock populations have grown rapidly in many parts of the Horn, and this growth has taken place on a fixed natural resource base. In many parts of the Horn, human population growth rates have been close to 3 percent a year, and fertility rates remain high. At these rates, the population will double every 25 to 30 years.

The growing number of humans and animals seems to be increasing vulnerability in some parts of this region. For example, pastoralists reported a 50 percent decline in median herd size over 1980–98 in northern Kenya, a region where human population growth was particularly rapid and land resources relatively constrained.[10] Other research suggests that the increasing competition over land in much of the region is largely a result of human population growth (partly owing to migrants from nonpastoralist areas).[11]

Policies and institutional factors may also be contributing to land fragmentation and reduced herd mobility. There have been significant efforts to expand irrigation in pastoralist areas, attempts to develop ranch-style livestock systems, and a consequent breakdown of community-based property right systems (through, for example, accelerated fencing of previously communal land). Underlying many of these trends are government policies and institutions that have typically done a poor job of protecting pastoralists' property rights.

Whatever its underlying causes, loss of mobility significantly weakens pastoralists' coping capacity. Areas with reduced mobility have been hardest hit in recent droughts in Kenya and Ethiopia. And more generally, sedentary farmers—typically ex-pastoralists—are poorer and more vulnerable than pastoralists, precisely because pastoralists can use mobility as a coping mechanism. Yet despite substantial evidence on the potential benefits of pastoralism in this kind of environment, central governments—which are often wary of mobile populations that regularly cross national borders unchecked—typically underappreciate the need for mobility.

In summary, the reasons why the region is seemingly more vulnerable are far more complex than is often understood. Yes, drought is a major factor, as is the oft-cited conflict in Somalia. Yet underlying these shocks are slower-moving stresses—such as the reduction of herd sizes and the loss of herd mobility—that have undermined the resilience of communities in the region. Identifying the deeper sources of these stresses is far from easy, but many informed observers agree that there is a vicious cycle at work related to interactions between population growth, local conflicts, land fragmentation, and reduced mobility.[12]

INCREASING RESILIENCE IN THE HORN OF AFRICA

Enhancing resilience in the Horn of Africa requires striking a balance between strengthening pastoralism—the region's traditional and still dominant economic activity—and promoting meaningful economic diversification. As it is, the region is already more diversified than is implied by the label "pastoralist." For example, in the Somali region of Ethiopia, almost 70 percent of households engage in livestock rearing, but a large share also produce crops (43.4 percent), firewood (17.0 percent), and charcoal (14.7 percent). A smaller number of households engage in various cottage industries such as mat making (6.3 percent), services (10.0 percent), trading (3.8 percent), and general labor or employment (2.4 percent).[13] Other regions show similar or even greater degrees of diversity.[14]

However, the most common alternative livelihoods generate low returns. Agro-pastoralism (a sector often composed of failed pastoralists) typically pays significantly less than pastoralism, whereas irrigated farming pays somewhat more and urban livelihoods pay much more (Table 1). Table 1 masks the fact that agro-pastoralists' rainfed farming is an extremely volatile livelihood, perhaps more so than pastoralism (since pastoralists can cope with drought through increased mobility). Moreover, the major secondary occupations of collecting and selling natural products, such as firewood and charcoal, pay the lowest of all.[15] These occupations are very much a negative coping strategy since they damage the environment and can impede pastoralism by removing the shrubs upon which livestock feed.

This evidence suggests that if pastoralist economies are to diversify, they should do so by expanding irrigated farming and increasing migration to urban areas. In the short to medium run, however, the basic issue is how many new entrants these alternative livelihoods can absorb. For example, a recent analysis estimated that additional irrigation investments in arid and semi-arid lowland regions in East Africa could profitably absorb a minimum of 3.2 percent of its rural population in 2020 and a maximum of 12.6 percent, depending on assumptions about viable farm size and irrigation costs (Table 2).[16] The percentage absorbed could be somewhat larger if the estimates include rainwater harvesting, which essentially offers seasonal irrigation opportunities.

But there are reasons to be cautious about irrigation potential. Dryland irrigation schemes in the region have often adopted inappropriate practices or technologies that have quickly become unsustainable and unprofitable. Irrigation schemes can also restrict pastoralists' access to key water points and dry-season grazing lands. And there are questions about how sustainable arid and semi-arid lowland irrigation is in the context of the lower rainfall predicted by climate change models, as well as about negative downstream impacts on neighboring communities.[17]

TABLE 1 Well-being by livelihood type in the Somali region of Ethiopia, 2005

Livelihood type	Average income[a]	Dietary diversity score[b]	Children immunized (%)	Adult literacy (%)
Pastoralism	217 (340)	4.3	24.4	13.7
Agro-pastoralism	97 (199)	3.4	19.6	11.4
Irrigated farming	254 (345)	3.9	35.4	12.5
Urban	1,081 (1,103)	6.8	49.4	49.9

Source: S. Devereux, *Vulnerable Livelihoods in Somali Region, Ethiopia,* Research Report No. 57 (Sussex, UK: Institute of Development Studies, 2006).

[a]Income is in 2005 birr per month. Figures in parentheses reflect average income when households with zero income are excluded from the calculation.

[b]Dietary diversity score is the number of different food groups consumed in the preceding 24 hours, with the indicator ranging from 0 to 13 food types.

TABLE 2 Profitably irrigable area in the arid and semi-arid lowlands of East African countries

Irrigation cost scenario	Estimated profitable increase in irrigated areas (hectares)	Projected rural population in 2020 (millions)	Percentage of six-person rural households that could work:	
			1 IRRIGATED HECTARE	0.5 IRRIGATED HECTARE
Low	522,850	50.0	6.3	12.6
Medium	320,689	50.0	3.9	7.8
High	266,085	50.0	3.2	6.4

Source: Authors' estimates based on data and methods described in D. Headey, A. S. Taffesse, and L. You, *Enhancing Resilience in the Horn of Africa: An Exploration into Alternative Investment Options,* IFPRI Discussion Paper 01176 (Washington, DC: International Food Policy Research Institute, 2012).

Migration and urbanization may seem more promising, but the main prerequisite for successful migration and urbanization is greater investment in education, since the alternative is usually low-return informal employment or crime. Currently, education outcomes in pastoralist areas are deplorable (see, for example, Figure 3 for Ethiopia). Yet there is tremendous potential for scaling up education. Improving education outcomes will not only facilitate economic diversification and migration, but also reduce fertility rates, empower women, and even improve local governance and community-based animal health and extension services. Moreover, the age distribution in pastoralist areas is heavily tilted toward the very young, so a big push on education could have major impacts even in the next 10 years or so. And the demand for education appears to have increased substantially among pastoralist communities.[18]

The age distribution in pastoralist areas is heavily tilted toward the very young, so a big push on education could have major impacts.

The mobility and isolation of pastoralists present challenges to scaling up education, but there are ways to overcome these, such as boarding schools, distance learning, and mobile schools, all of which have strengths and weaknesses. Mobile schools, for example, are compatible with pastoralism but tend to have poor-quality teachers and limited resources. Boarding schools are potentially more attractive—and could be linked with school feeding programs and health interventions—but their promotion needs to be compatible with cultural and religious norms.

Health and nutrition interventions are also important ways of building up human capital. Although chronic malnutrition (reflected in stunted growth) is relatively low in pastoralist areas, acute malnutrition (reflected in wasting) is generally extremely high[19] because of high exposure to drought and acute food shortages, as well as extremely poor access to health services (recall the low immunization rates in Table 1). Adequate health and nutrition are not only important in their own right, but also necessary for improving school attendance and performance, so a broader strategy for human capital development in the region will yield high dividends. And health interventions are at the top of pastoralists' own development priorities.[20]

Although diversifying the region's economies is pivotal, it is also important to make pastoralism more profitable and resilient for two reasons. First, diversification strategies take time to bear fruit and have limited capacity to absorb more people in the near future. Second, mobile livestock rearing has a comparative advantage in a land-abundant region with volatile rainfall. Indeed, in some ways livestock trade in the region has been doing very well. During the 1990s, Somalia's livestock exports to Kenya doubled.[21]

FIGURE 3 Literacy status in Ethiopia by pastoralist and nonpastoralist districts

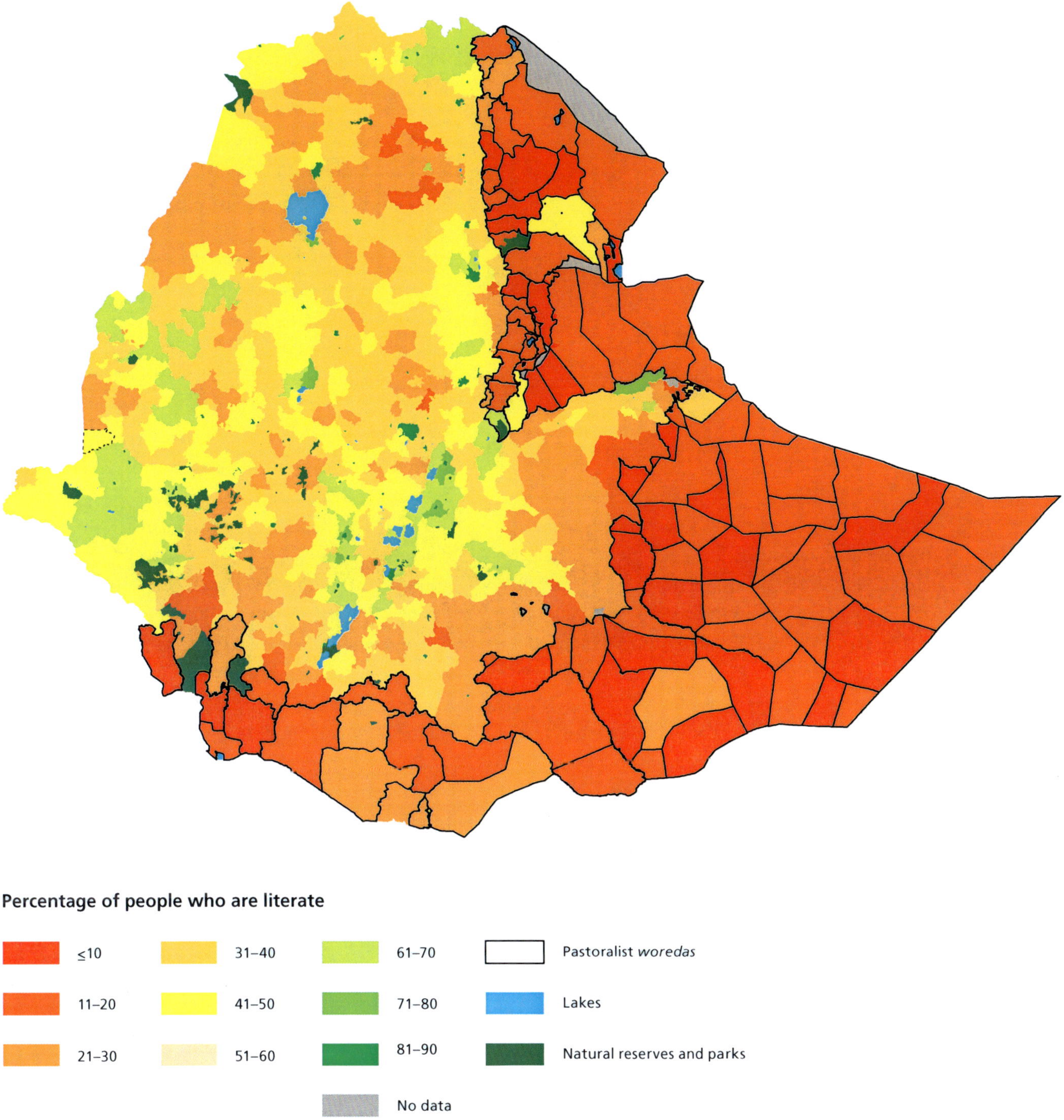

Source: Author's construction based on data from the Central Statistical Agency (CSA) of Ethiopia. **Note:** A *woreda* is a district.

In the 2000s, Ethiopia's formal livestock exports rose from just US$8 million in 2004 to just over US$200 million in 2010.[22] The vast majority of Ethiopia's livestock exports—90 to 95 percent—are sourced from pastoralist areas.[23] These positive trends come on the back of strong overseas demand and higher prices.

Improving market access and integration would allow pastoralists to buy and sell livestock before a drought rather than lose their herds to drought-induced mortality.

But how can the livestock sector be further developed in a manner that is pro-poor? Efforts to develop pastoralism have focused on commercialization—that is, promoting greater engagement with markets—and improved drought management, but there is little evidence that commercialization interventions in the Horn have benefited the poor (partly because the evidence base is weak).[24] Moreover, there are signs that inequality in pastoralist areas may be increasing. Large herders have increasingly engaged in overseas markets and coped relatively well with drought, while poor herders have often failed to sell their livestock before drought and subsequently lost most of their herds. These ex-pastoralists are therefore forced to work as hired herders or in agro-pastoralism or other low-return activities.

Improving market access and integration would allow pastoralists to buy and sell livestock before a drought rather than lose their herds to drought-induced mortality. To achieve this, road infrastructure is obviously a high priority, bearing in mind the need to keep such investments strategic and cost-effective in low-population-density areas. Information systems could be better developed to provide weather forecasts and early warnings, livestock prices, and other pertinent information for both traders and pastoralists (such as advance notice of livestock auctions). Cellular phones have been used to disseminate early warnings and price data, but more could be done to make markets more competitive. In smaller livestock markets, traders often appear to have more bargaining power than pastoralists, who can ill afford to trek unsold animals back to their grazing lands. One solution could be to establish specific market days and shift to an auction-based system.[25] In theory, these relatively simple institutional changes should increase the prices received by pastoralists and even promote broader commercialization of the sector.

Changes related to trade and animal health are also needed to render the pastoralist sector more viable. Most livestock exports in the region remain informal, partly because of onerous regulations and poor customs infrastructure.[26] In Ethiopia, emergency animal health interventions typically have low returns since the main constraints during drought are food and water.[27] However, improving animal health in normal times remains extremely important both for protecting and building up this key household asset, and for preventing the spread of diseases and subsequent bans on livestock exports, which can impose a huge economic cost on the region.

Better management and regulation of land and water resources will also be critical. In addition to suffering from the effects of conflict over land and water, herders have felt the negative impacts of irrigation schemes, "land grabs," and the generally poor protection of community property rights. These land policies are not only unjust, but also inefficient because they inhibit the performance of the livestock sector and interfere with the principal mechanism pastoralists use to cope with drought. This record reflects the pervasive marginalization of pastoralist communities by national governments. In recent years, though, a number of interventions and institutions have been developed to redress this marginalization,[28] and significant advances have been made in resolving local conflicts, including regulation of grazing and water resources.[29]

Finally, the need for evidence-based strategies is an issue that pervades every aspect of development strategies in the region. Even the most experienced researchers on the Horn of Africa acknowledge the lack of information on what works and what does not. Developing evidenced-based strategies means trying out technological and institutional innovations to cope with the region's problems—and then rigorously evaluating those innovations. Relevant ideas and technologies can be imported from extensive livestock systems elsewhere in the world, such as Africa, Australia, Central Asia and China, the Middle East, and North America. New technologies could include greater use of cellular phones for market and early warning information, satellite-based weather information on rainfall and pasture availability, index-based livestock insurance, and improved seeds. Institutional innovations might include improved regulation of water points, creation of livestock corridors (especially where irrigation schemes are present), strategic investments in infrastructure with stronger links to livestock centers, value-chain interventions (such as fattening of livestock), mobile schools and clinics, and public–private partnerships to encourage private investment in the region. All of these schemes could help mitigate the disadvantages of distance and the vagaries of the climate, but figuring out what works and what does not will require greater experimentation and more rigorous evaluation.

MOVING FORWARD

Major climatic shocks in the Horn of Africa are inevitable, but human vulnerability to these shocks is not. Promoting social, economic, and ecological transformation in the region could build up resilience to these shocks and mitigate the slower-moving stresses that also undermine progress in the Horn. Achieving that resilience requires investing more in both livestock and nonlivestock sectors, rapidly expanding infrastructure and human capital, making synergistic improvements in disaster risk management and development interventions, and improving governance and conflict resolution efforts. The precise instruments for achieving these outcomes are less obvious, but they must inevitably be the product of innovation, experimentation, and—not least—political commitment. ■

Chapter 4

CLIMATE CHANGE AND AGRICULTURE

Modest Advances, Stark New Evidence

Gerald C. Nelson and Tolulope Olofinbiyi, IFPRI

THE YEAR 2011 BROUGHT BOTH GOOD AND bad news about climate change and agriculture. The good news is that after initial steps toward rebuilding confidence in the United Nations' climate change negotiations were taken in Cancun in December 2010, further progress occurred in Durban in 2011. And outside the formal negotiations process, many countries have begun to implement their own mechanisms to reduce greenhouse gas emissions and to adapt to some climate changes that increasingly seem inevitable.

The bad news includes growing evidence that climate change has already affected agricultural productivity[1] and will put increasing pressure on agriculture in the coming decades. Recordbreaking extreme weather events around the world in 2011 offered a glimpse of the challenges climate change will bring. Farmers worldwide will need to adapt to higher temperatures and shifting precipitation patterns. In addition, climate variability will likely cut into global food production, exacerbating the existing problems of poverty, food insecurity, and malnutrition. In addition, after declining in the wake of the global financial crisis, greenhouse gas emissions are once again rising rapidly, making the climate change challenge to food security much greater.

REBUILDING CONFIDENCE IN INTERNATIONAL NEGOTIATIONS, SLOWLY

Delegates to the United Nations Framework Convention on Climate Change arrived in Copenhagen in December 2009 with great optimism that an agreement could be reached to reduce greenhouse gas emissions and provide financial support to help developing countries adapt to climate change. Most of the world's leaders were scheduled to arrive in the second week, when they would undertake the final political negotiations needed to close the deal. As the second week arrived, however, a deal was nowhere

BOX 5

Better Tools for Tackling Climate Change

Bruce Campbell, CGIAR Research Program on Climate Change, Agriculture and Food Security

Farmers and policymakers in developing countries need support in their struggle to adjust to global changes in climate. They must have evidence to weigh the pros and cons of different strategies and policies. Providing that support through research-based evidence is the goal of the CGIAR Research Program on Climate Change, Agriculture and Food Security (CCAFS), which was in its first full year of operation in 2011. Inevitably, adjusting to climate change will mean making complex tradeoffs among food security, livelihoods, and the environment both at grassroots levels and in the corridors of power. Sound research will help policymakers, farmers, and others affected by climate change understand the implications of their decisions when making difficult compromises.

Progressive climate change threatens farmers in developing countries, who must adapt their farming practices to the changes ahead in order to survive. Studies by CCAFS Program scientists published in the book *Crop Adaptation to Climate Change* describe how climate change could threaten production of important food crops such as potatoes, beans, bananas, and cassava—and how specific adaptation strategies such as new plant breeds could neutralize or at least significantly lessen the impact.[1] As part of the Program's work on adaptation through managing climate risks, researchers have been training farmers on interpreting seasonal climate forecasts in East and West Africa. The Program's researchers also produced a study on "Mapping Hotspots of Climate Change and Food Insecurity in the Global Tropics" to identify food insecure areas most vulnerable to the impacts of future climate change, across the priority regions for the CGIAR centers.[2] A workshop and paper series examined how new institutions, property rights arrangements, and agricultural technologies can improve livelihoods and reduce greenhouse gas emissions.[3] As part of its ongoing work on gender, the Program issued grants to six female scientists working in Program target regions to study the links among gender, climate change, agriculture, and food security.[4]

Climate change was on the agenda of many international institutions in 2011, including two large conferences: the United Nations Framework Convention on Climate Change in Durban, South Africa, and the International Conference on Climate Change and Food Security in Beijing, China. Key agricultural organizations (including the Program, the World Bank, and IFAD) coordinated Agriculture and Rural Development Day, a parallel event at the UN conference that focused on galvanizing international support for a new work program on agricultural climate change adaptation and mitigation.

The Program's work in 2011 focused on taking stock and developing the relevant research strategies that will have the greatest impact. For example, researchers implemented a baseline food security and climate adaptation survey covering more than 5,000 households in more than 250 villages across 36 sites in 12 countries in East and West Africa and South Asia.[5] The data gathered in the past year will help provide decisionmakers with evidence-based results and useful tools for designing and testing approaches to adaptation and mitigation. When their work is completed, researchers will be able to report whether certain techniques were successful. The Program has also created the Climate Change Adaptation and Mitigation Knowledge Network, an information service and a key tool for practitioners, donors, policymakers, and researchers interested in food security and climate change. The Network is a map-based online platform that brings climate, agriculture, and socioeconomic information together, and uses multimedia to share stories of farmers living at research sites across the tropics.

Climate change affects agriculture and food security in a variety of ways, so choosing the best mitigation and adaptation techniques requires thorough research. The CGIAR Research Program on Climate Change, Agriculture and Food Security contributed significant evidence in 2011, but this is only the beginning of an undertaking that must reach beyond a single research program to match climate change's complexity with its own breadth and depth.

in sight. The impending arrival of several political leaders on Friday, the official closing day, meant that negotiators needed to engage in extremely high-level talks to develop some kind of agreement. After several sleepless nights, the negotiators took note of a document called the Copenhagen Accord, emanating from several high-level meetings. This accord enshrined the goal of keeping the average temperature rise to 2°C and pledged US$10 billion a year from developed countries over the next three years, rising to US$100 billion a year by 2020, to help poor countries adapt to climate change. None of these commitments, however, were binding, and it is unclear that any have been or will be met.

Although the Copenhagen negotiations were unsuccessful overall, they marked the start of a push to formally include agriculture in the negotiation outcomes, with the first Agriculture and Rural Development Day providing a convening venue for those concerned about the challenges to agriculture from climate change. Buttons bearing the slogan "No agriculture, no deal!" made their first appearance and have become an increasingly common sight at United Nations negotiations.

At the following year's convention, delegates arrived in Cancun in late November 2010 with greatly lowered expectations and no plans for the attendance of large numbers of heads of state. With low expectations as a starting point, the eventual outcomes were substantial. Negotiators approved a large number of documents, collectively called the Cancun Accord. Important elements included the reaffirmation of the 2°C temperature increase target, improved reporting requirements, and the start of a process to design a Green Climate Fund. The second Agriculture and Rural Development Day was held, with delegates pushing negotiators to formally include agriculture in any outcomes and calling for approval of an official work program on agriculture. Unfortunately, the work program was lost in the final days of the negotiations when it got caught up in disagreements about whether individual sectors should be singled out and a perception by some negotiators that a work program would focus only on mitigation and ignore adaptation.

The 2011 Durban round of negotiations was widely perceived as particularly important for Africa, because a successful outcome would reflect well on the continent and because Africa is likely to be seriously affected by climate change[2] and so would have much to gain from a successful outcome. One of the key challenges was the pending expiration of the Kyoto Protocol in 2012. The protocol set binding targets for emissions for most developed countries (the United States was not a signatory and therefore not a party to the emissions reduction commitments). Without an extension of the protocol, countries would no longer be legally bound to reduce their emissions.

Although the Copenhagen negotiations were unsuccessful overall, they marked the start of a push to formally include agriculture in the negotiation outcomes.

With African agriculture especially threatened by climate change, a major push was made to persuade the negotiators to include an official work program on agriculture. The activities of the third Agriculture and Rural Development Day were organized around this goal, and major public figures, including former UN secretary general Kofi Annan and Jacob Zuma, president of South Africa, pressed the negotiators to approve the work program on agriculture.

By the last Friday of the negotiations, it was unclear whether anything would be achieved. But negotiators agreed to continue their work and extended their string of sleepless nights, ultimately finishing Sunday morning. The outcome of this effort is called the Durban Platform for Enhanced Action. A key element of the platform is that all Kyoto signatories plus the United States agreed to forge a treaty by 2015 that would bring all countries, developed and developing, under a legally binding agreement by 2020. For the first time, China and India, two of the world's largest emitters of greenhouse gases, agreed to this principle.

The Green Climate Fund was formally established and awaits contributions from member countries. A second commitment period for the Kyoto Protocol was approved with the reduction targets to be determined at a meeting in 2012. However, Canada, one of the important countries to sign on to the first commitment period, announced it would not join the second period, and other developed countries have suggested that they may follow suit. Finally, although the negotiators did not adopt an official work program on agriculture, they did approve a process for developing a work program for approval at the next negotiating session, to be held in Qatar beginning in late November 2012.

GOING IT ALONE: INCREASING PROGRESS OUTSIDE FORMAL NEGOTIATIONS

Although progress remains extremely slow in the official negotiating process, countries around the world are beginning to devote substantial resources to agricultural adaptation and mitigation activities that could have high payoffs today and lead to increased resilience tomorrow.

India

India continues to launch adaptation programs at both the national and state levels. Adaptation activities in different areas of the country include efforts to improve and diversify crops, conserve soils, develop watersheds, manage irrigation water, and improve disaster management through, for example, drought and flood proofing. An example of one of these activities is a community watershed project designed to explore low-cost water conservation solutions to improve crop yields in the face of drought in the community of Kothapally in Andhra Pradesh, India. This long-term project, developed by the International Crops Research Institute for the Semi-Arid Tropics at the request of the government of Andhra Pradesh, was designed with and managed by the community. A recent impact study showed that the project's water management practices improved infiltration and water-holding capacity of the soil, increasing water availability by 10 to 30 percent and raising crop yields. The study suggests that implementing agricultural water interventions on a large scale can significantly raise agricultural productivity and increase farmer livelihoods.[3] Still, more needs to be done to increase the adaptation of agricultural systems in India because climate change threats to productivity are expected to grow. Additional adaptation strategies—such as more efficient use of water, promotion of eco-friendly technologies, shifts in cropping patterns, and agricultural insurance—should be considered, and adaptation and mitigation programs should be mainstreamed into national agricultural strategies.

China

The government of China, which launched a national Climate Change Program in 2007, has explored several strategies and activities to help the agricultural sector adapt to climate change.[4] Some of the efforts include improvements to agricultural infrastructure and increased investment in research and development of new technologies. To improve agricultural infrastructure, the government has, for example, accelerated the construction of water-saving irrigation projects. As part of special funding arrangements established for climate change adaptation, China has invested in new technologies such as cloud seeding to promote reliable rainfall. The government has also increased pilot projects on different types of insurance policies. Because these efforts were initiated recently, their effectiveness has not been assessed and documented.

Kenya

A recent study of four agroecological zones in Kenya shows there are win-win-win agricultural practices that can pay off in terms of adaptation, mitigation, and profitability. For example, when poor smallholder producers use sustainable agricultural management practices, they not only increase their resilience to climate change and variability, but also contribute directly to reducing greenhouse gas emissions and increasing agricultural productivity and profitability. In particular, soil nutrient management—applying combinations of inorganic fertilizer, mulch, and manure—is

FIGURE 1 Carbon dioxide emissions in developed and developing countries, 1990–2010 (PgC)

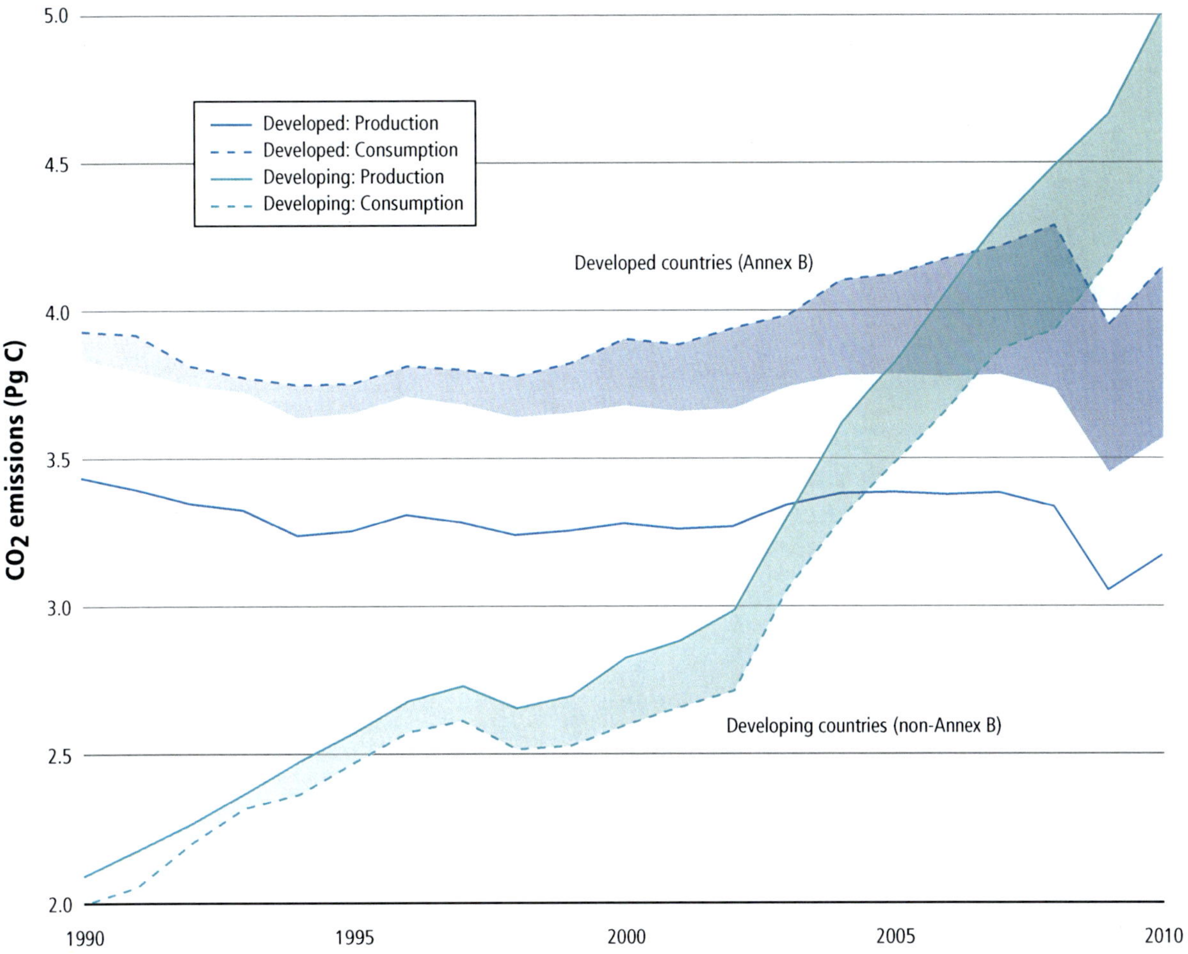

Source: Figure 2 in G. P. Peters, G. Marland, C. Le Quéré, T. Boden, J. G. Canadell, and M. R. Raupach, "Rapid Growth in CO2 Emissions after the 2008–2009 Global Financial Crisis," *Nature Climate Change* 2, no. 1 (2012): 2–4, doi:10.1038/nclimate1332. **Note:** Carbon dioxide emissions are for fossil fuels.

shown to enhance crop yields, soil carbon stocks, and incomes from agricultural production. Similarly, introducing improved feeds for dairy cattle decreases methane emissions per liter of milk and increases profitability in most parts of Kenya. These improved practices can allow livestock producers to reduce the numbers of livestock and lower overall emissions while increasing food production and food security. In the arid zone, farmers can use irrigation and soil and water conservation methods to maximize soil carbon and agricultural profits. These win-win-win actions, however, have yet to be strategically exploited. To do so will require building capacity among decisionmakers at the national level to ensure that they explicitly include climate change adaptation and mitigation in their agricultural productivity and food security strategies and policies. It will also require improving farmers' access to financial resources, such as voluntary carbon markets and adaptation and mitigation funds.[5]

NEW EVIDENCE ON THE THREATS TO AGRICULTURE FROM CLIMATE CHANGE

To illustrate the challenges in mitigating greenhouse gas emissions, Figure 1 contrasts trends in domestic carbon dioxide emissions in developed and developing countries. For developed countries, emissions from domestic production have

Recent research strongly suggests that rising temperatures and accompanying changes in precipitation have already had observable effects on agriculture.

remained relatively flat for the past 30 years, with a large dip in 2008 as the global economic crisis reduced economic activity. Emissions from consumption have continued to grow in the form of emissions associated with imports produced in developing countries.

In stark contrast, developing-country emissions have grown dramatically, surpassing those from developed countries by the late 2000s. The green shaded area indicates that a significant and growing portion of developing-country emissions are from production for export to developed countries, but emissions from domestic consumption have surpassed those from developed-country consumption. This situation makes it increasingly obvious that developing countries should not be excluded from national commitments to reduce emissions, as they were in the Kyoto Protocol. As incomes in developing countries grow, they must pursue low-emissions development strategies.

Climate scientists are increasingly confident of the link between recent anthropogenic emissions of greenhouse gases and rising temperatures such as those experienced in Iowa, in the heart of the US Corn Belt. Researchers have demonstrated the threat to US maize production from higher temperatures based on a detailed statistical analysis of actual maize yields from 1950 to 2005.[6] Their findings show that as average growing season temperatures rise to 28°C, yields are relatively little affected. However, once temperatures surpass the threshold of about 30°C, yields drop precipitously. Another study using extensive maize experimental data from International Maize and Wheat Improvement Center trials in Sub-Saharan Africa had broadly similar results.[7] Under optimal management, when the mean growing season temperature is less than 22°C, a 1°C increase in temperature has a small but positive effect on yields. But as the average growing season temperature exceeds 25°C, the effect becomes negative, causing roughly a 30 percent decline in yields. And during

FIGURE 2 Change in growing season temperature, 1980–2008

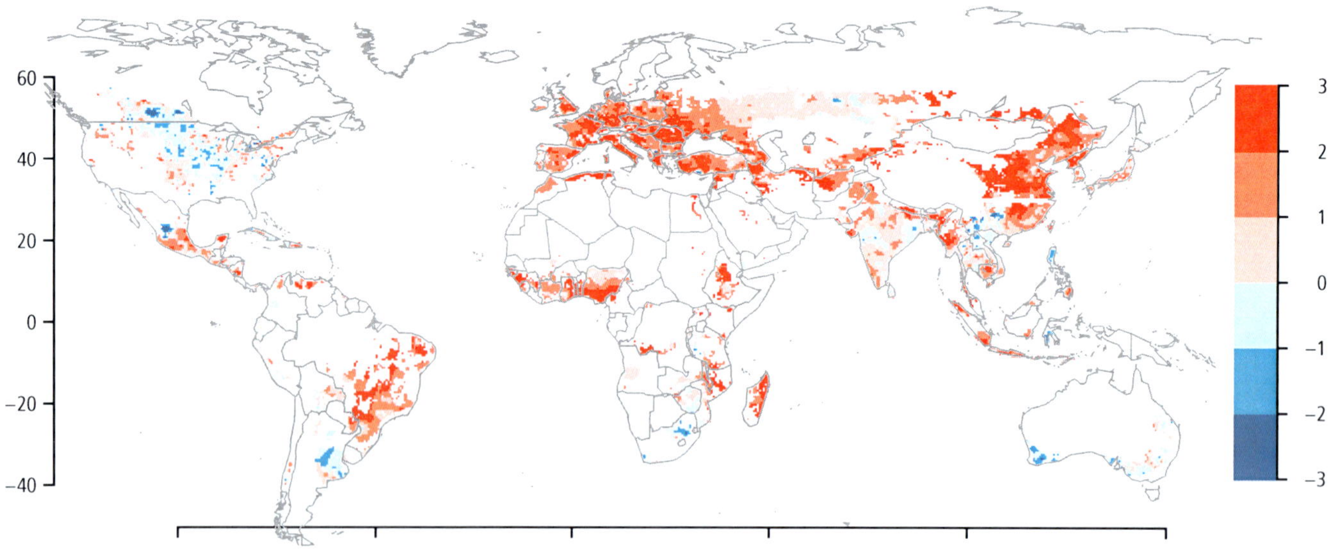

Source: Figure 1 in D. B. Lobell, W. Schlenker, and J. Costa-Roberts, "Climate Trends and Global Crop Production since 1980," *Science* 333, no. 6042 (2011): 616–620, doi:10.1126/science.1204531.

FIGURE 3 Estimated net impact of climate trends for 1980–2008 on crop yields, divided by the overall yield trend

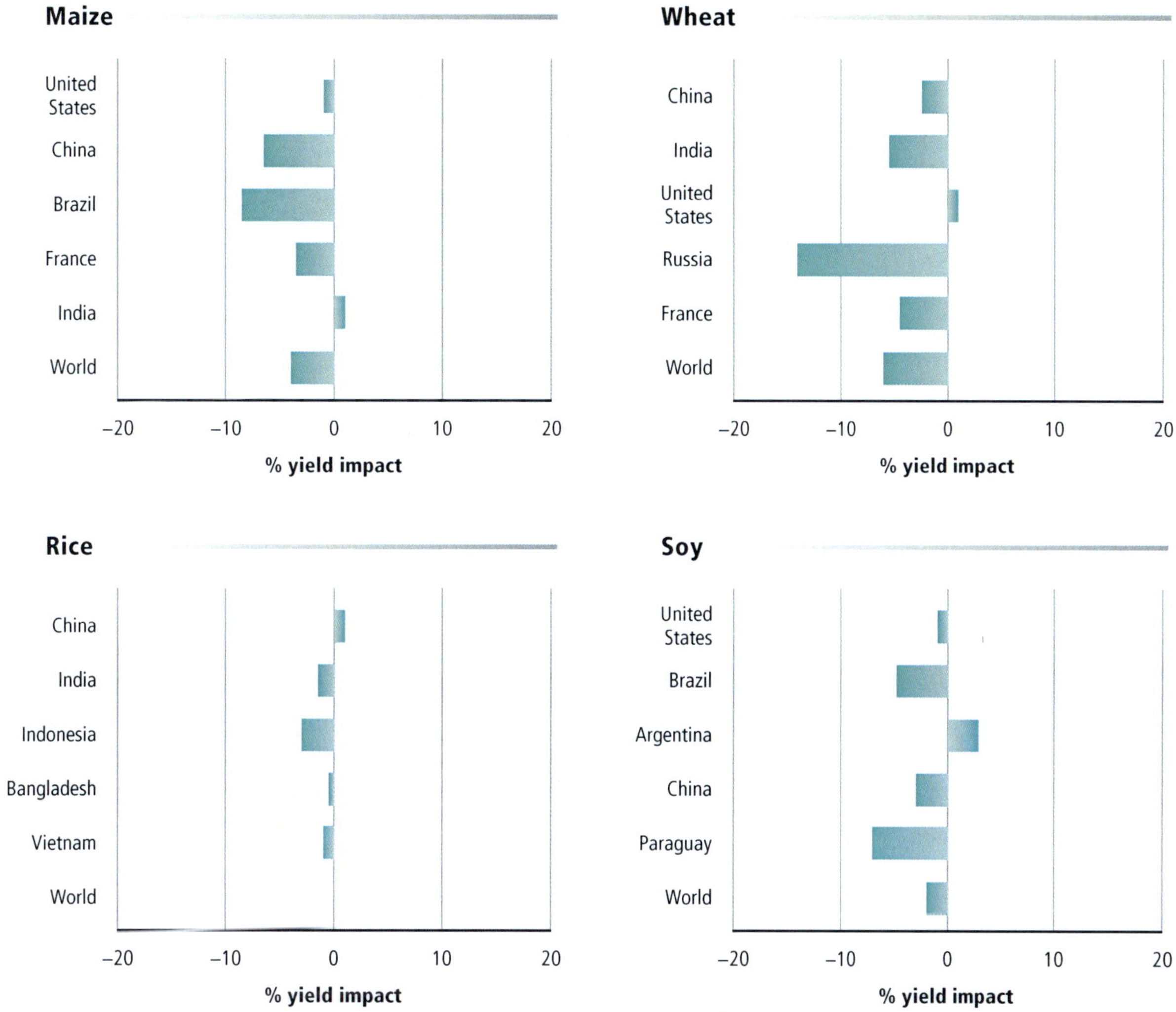

Source: Figure 3 in D. B. Lobell, W. Schlenker, and J. Costa-Roberts, "Climate Trends and Global Crop Production since 1980," *Science* 333, no. 6042 (2011): 616–620, doi:10.1126/science.1204531.

a drought, the yield declines begin at lower temperatures and can be greater than 40 percent.

Other recent research strongly suggests that rising temperatures in the second half of the 20th century and early years of the 21st century, and accompanying changes in precipitation, have already had observable effects on agriculture. Although growing season temperature changed only slightly in North America from 1980 to 2008, it increased dramatically in other parts of the world, particularly China and Europe (Figure 2).

The consequence of the differing temperature increases can clearly be seen in the changes in yields (Figure 3). For maize, climate change had essentially no effect on US yield trends, whereas it substantially slowed yield growth in Brazil, China, and France. In some countries, however, regional crop production has benefited from higher temperatures. The growing area has shifted northward for maize in the United States, rice in China, and wheat in Russia.

THE CHALLENGE TO FOOD SECURITY

The precise temperature and precipitation changes that climate change will bring, as well as the

FIGURE 4 Scenarios of climate change and food security

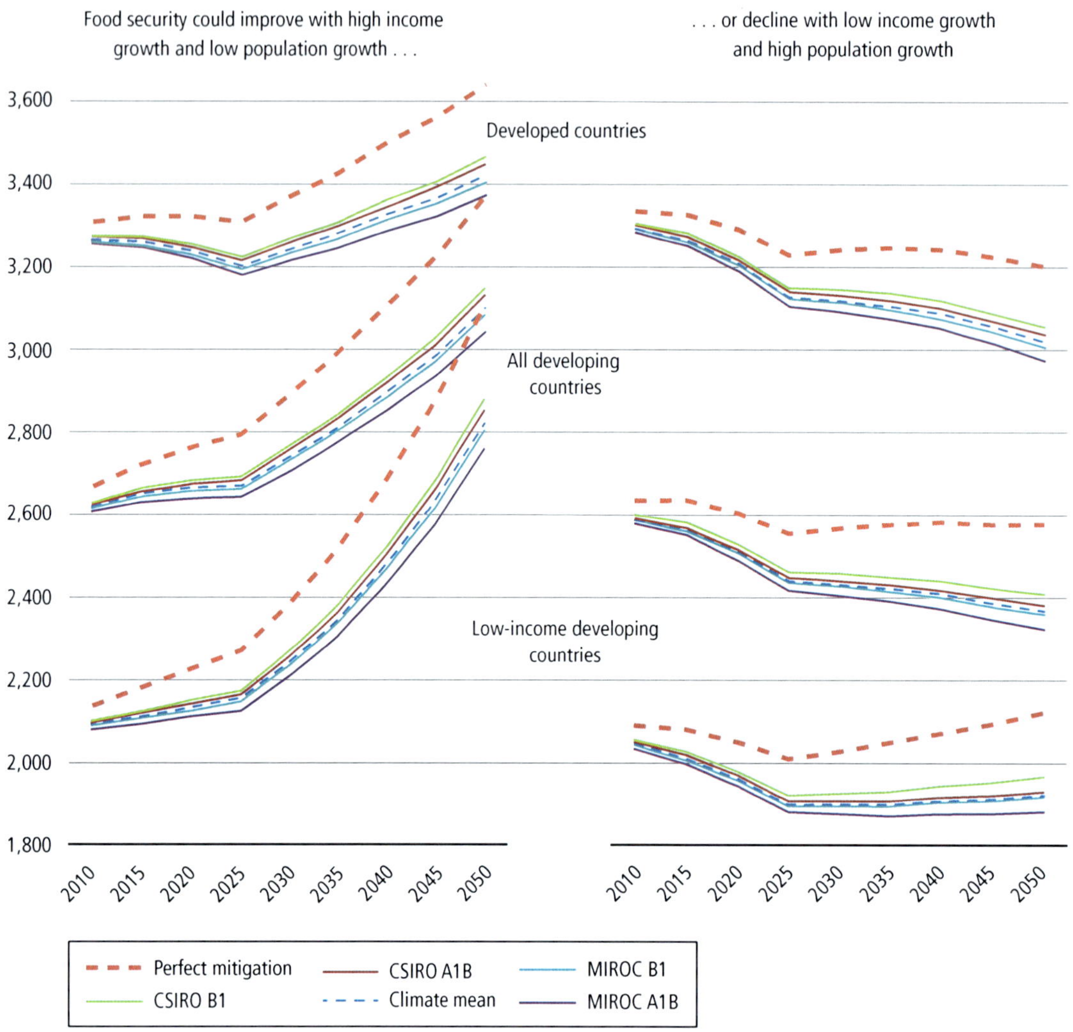

Source: Figure 3.1 in G. C. Nelson, M. W. Rosegrant, A. Palazzo, I. Gray, C. Ingersoll, R. Robertson, S. Tokgoz, et al., *Food Security, Farming, and Climate Change to 2050: Scenarios, Results, Policy Options,* IFPRI Research Monograph (Washington, DC: International Food Policy Research Institute, 2010). **Notes:** The dashed red lines show calorie availability in an unrealistic scenario with perfect mitigation (that is, all emissions stop today and the existing momentum in the climate system is also stopped). The solid lines of various colors reflect outcomes with plausible climate results from two general circulation models, each with two scenarios, from the *Intergovernmental Panel on Climate Change's Special Report on Emissions Scenarios.* The shift in values in 2025 reflects the assumption that maize-based ethanol will be replaced with cellulosic ethanol around 2025. This change will reduce the demand for maize, lower its price, and make more calories available for human consumption.

context within which they will take place, are still uncertain. Consequently, to design policies that can protect populations vulnerable to climate change and increase the likelihood of achieving sustainable food security, it is critical to understand the impacts of climate change under different scenarios. Figure 4 shows how climate change will likely challenge food security. It reports average calorie availability per person per day—an imperfect measure of food availability—under a range of climate change scenarios and two overarching scenarios of the development context. The optimistic scenario reflects high income growth and low population growth, representing a situation of sustainable development. The pessimistic scenario consists of low income growth and high population growth.

Three messages stand out from the results in Figure 4. First, sustainable development, embodied in the optimistic scenario, is key to improving the well-being of the poorest. With sustainable development, calorie availability improves dramatically in the 40 countries with the lowest income today. Second, climate change significantly reduces calorie availability around the world, as shown by the gap between a scenario of perfect mitigation and the rest of the climate change scenarios. Finally, although the different climate change scenarios have substantially different consequences for agricultural productivity, the final outcomes for calorie availability are similar. This result is due to dramatic differences in trade flows in the different scenarios. Thus, relatively open international trade will be a crucial part of adapting to climate change.

Rapidly increasing greenhouse gas emissions, especially in developing countries, combined with growing evidence of negative climate change effects on agriculture, the likelihood of nonlinear effects of temperature on yields, and hints of the added burden of more frequent extreme weather events suggest an extremely serious challenge for sustainable food security. Renewed efforts to reduce greenhouse gas emissions and redoubled efforts to develop crop and livestock varieties, along with management systems that remain productive with higher temperatures and more extremes in heat and precipitation, are crucial. In 2011, the body of evidence on the threat to food security from climate change became increasingly robust. The challenge is to find the resources to address the problems before they overwhelm us. ■

BIOFUELS, ENVIRONMENT, AND FOOD

The Story Gets More Complicated

David Laborde and Siwa Msangi, IFPRI

POLICY DISCUSSIONS OVER BIOFUELS IN 2011 reflected the increasing complexity of the issue over the past decade. Originally promoted as a way of decreasing dependence on fossil fuels and avoiding the carbon emissions generated by them, biofuel production has now been widely recognized to have strong links to agricultural markets and even land-use patterns. The first-generation biofuels currently in commercial use—biodiesel made from vegetable oil and ethanol made from sugarcane or maize—have the strongest links to agricultural markets and land use, although there are important by-products that can be used as feed in order to offset these effects. Given the areas of scientific uncertainty that still remain over the impact of biofuels on food security and the environment, decisionmaking has become complex and, in some cases, contentious.

Nearly a dozen international institutions came together in 2011 to issue a joint report that addressed the issue of biofuels and food prices and called for eliminating distortive biofuel policies, especially where the environmental benefits are not as high as expected.[1] The Group of 20 (G20) also raised the issue of biofuels in 2011 as part of its overall concern with food security. The G20 countries recognized the need to examine the role of biofuels in food price volatility and to adjust biofuel mandates when market situations warrant interventions. They did not, however, make more definitive statements about biofuels and their links to food prices because of disagreements between large producers (like Brazil) and net food importers (like China) on the importance of these links.[2] The role of policy support to domestic biofuels sectors—in the form of tax credits, subsidies, and tariffs against imported ethanol (for the United States and the European Union)—remains a concern for key stakeholders.

Indeed, key countries display different social preferences in handling the delicate issue of food–fuel links depending on the local dynamics of agricultural demand and supply. In Brazil, the flexibility of sugar-ethanol mills allows producers to shift easily between ethanol and sugar production based on prevailing market conditions for food (sugar) and fuel (ethanol/gasoline) and provides them with constant, year-round outputs in their supply chain. Since 2008 Brazil has reduced its exports of ethanol to the world market, in part because of rising US production and exports, increased demand for sugar from large consumers like India, and high and uncompetitive prices in its domestic ethanol market.[3] As a result, in 2010 and 2011 Brazil found itself importing ethanol from the United States, boosting US biofuel revenues, and making the US tariff against Brazilian ethanol imports inconsequential.

China has backed off of aggressive expansion of biofuel production in the past five years because of concerns about domestic grain markets and prices. China's biofuel production started rapidly with the building of four state-owned ethanol plants in 2001. By 2007, it had produced a total of 1.35 million tons of ethanol, placing it third in the world. At that point, this rapid rise was halted and the use of cereals in biofuel production was capped.[4]

Despite some attention to the impacts of biofuels on food security, much of the policy discussion over biofuels in 2011 focused on environmental concerns.[5] International biofuel markets are dominated by the European Union and the United States, the largest consumers and producers of biodiesel and ethanol, respectively. While neither adopted major policy changes in 2011, the year was still one of intense debate, paving the way for potentially important decisions in 2012. Although the policy debate focused on the environment, any decisions made regarding biofuel production will have implications for global food markets, given the volume of crop-based feedstocks that are converted annually.

EUROPEAN UNION

In the European Union the consumption of biofuels is a key component of a decision to reduce greenhouse gas emissions from the transport sector by replacing fossil fuels with renewable energy. In 2003 a European Union directive set a target of 5.75 percent for renewable-energy use in the transport sector by 2010. In 2009 the European Union adopted the Renewable Energy Directive, which has a target of 10 percent by 2020. Although renewable energy can include electricity, hydrogen, or second-generation biofuels (that is, ethanol and biodiesel made from nonfood feedstocks such as agricultural residues and switchgrass), the main mechanism for meeting this target is and will remain first-generation biofuels.

The directive also established environmental sustainability criteria for biofuels, including a minimum rate of direct greenhouse gas emission savings (35 percent in 2009, rising to 50 percent in 2017) and restrictions on the types of land that may be converted to production of biofuel feedstock crops. This restriction covers direct land-use changes only. The revised Fuel Quality Directive, adopted at the same time as the Renewable Energy Directive, is more technical, includes identical sustainability criteria, and targets a 6 percent reduction in greenhouse gas emissions from transport fuels by 2020.

Because reducing emissions is officially the only goal of Europe's biofuel policy, policymakers set a high priority on getting a correct assessment of the greenhouse gas balance sheet of biofuels. Such an assessment would account for the diversion of land use in biofuel production, which can reverse the conclusion about biofuel's capacity to reduce carbon emissions.[6] Land-use changes occur when farmers, domestically and abroad, replace production historically dedicated to food and feed with production of biofuel crops or when they convert natural land to cropland. These land-use changes are considered *direct* if farmers convert natural land cover directly to cropland for growing the feedstock crop used in biofuels within the biofuel-producing country. On the other hand, land-use changes are considered *indirect* if changes in market prices cause another crop to expand into natural land cover or if a reduction in exports from the biofuel-producing country (such as maize from the United States) causes farmers in other countries to convert natural land to cropland to expand production of those (or other) crops. Given the complex nature of domestic and international market linkages, indirect land use changes are much harder to verify and observe than direct land-use conversions. Therefore, in 2009 the European Council (representing the governments of member states) and Parliament asked the European Commission to examine the question of indirect land-use change, including possible measures to avoid it, and report back on the issue by the end of 2010.

The Commission then launched four studies to examine indirect land-use change issues. One study, conducted by the International Food Policy Research Institute (IFPRI), analyzed the impact of the European biofuels mandate and possible changes in Europe's biofuel trade policies on global agricultural production and the environmental performance of the European biofuel policy, as spelled out in the Renewable Energy Directive.[7] The report suggested that indirect land-use change was a valid concern but that there was a high degree of uncertainty regarding its magnitude. Following these investigations and public consultation, in December 2010 the Commission published a report acknowledging that indirect land-use change can reduce the greenhouse gas emissions savings associated with biofuels. Because of the many uncertainties, however, the Commission did not deliver a clear recommendation about whether and how measurement of indirect land-use change should be included in the legislative framework. The Commission announced that new research would be conducted and that an impact assessment report would propose several policy options.

During 2011 the discussions became more intense. On the one hand, biofuel producers disagreed with the concept of indirect land-use change and claimed that even the debate and uncertainty about future legislation deters investments and is costly to Europe's economy and climate change strategy. On the other hand, many members of the European scientific community and observers from the United States asked the Commission to reconsider its position regarding biofuels and urged it not to make emissions accounting mistakes regarding biofuels. Nongovernmental and environmental groups actively highlighted the social risks linked to biofuels (such as "land grabbing" and competition between food and fuel uses) as well as the environmental risks (such as increased emissions). Although the European Commission had not yet released its impact evaluation report by the end of 2011, it did release a new modeling exercise conducted by IFPRI on the land-use issue in October.[8] Reflecting the fact that the merits of first-generation biofuels are highly disputed, the Commission also stated that it would no longer support biofuel projects in its overseas development policies.

UNITED STATES

In the United States, discussions of biofuel policy take place at two levels. At the federal level, the US Environmental Protection Agency regulates biofuel blending through the Renewable Fuel Standard. At the state level, some ambitious states have set up their own biofuel policies (such as California's Low Carbon Fuel Standard), seeking to improve upon the Renewable Fuel Standard in terms of environmental performance.

In 2011 there were a number of policy discussions at the federal level about whether the biofuel tax credit—called the Volumetric Ethanol Excise Tax Credit—should be repealed. In an atmosphere of increasing fiscal austerity within the United States, an unusual alliance of fiscal and social conservatives and environmentally minded opponents of biofuels emerged around the issue of repealing the tax credit. Researchers have pointed out the welfare and efficiency losses that result when such a tax credit is combined with a blending mandate, which is part of the federal Renewable Fuel Standard policy.[9] Food security concerns have been raised over the effects of tax credits and subsidies on biofuel production and, in turn, on the level and stability of agricultural and food prices.[10] Some have pointed out, however, that energy prices were a stronger driver of past growth in biofuel production than tax credits alone.[11]

What if the current Renewable Fuel Standard were replaced with (or complemented by) a policy focused on lowering the carbon intensity of fuel, such as California's Low Carbon Fuel Standard? The essential difference between the two policies involves the incentives for biofuel producers and blenders. Because the Renewable Fuel Standard awards credits based on the production and blending of ethanol and biodiesel, it rewards biofuel production regardless of whether and how much that production reduces carbon emissions. In contrast, the Low Carbon Fuel Standard allocates credits to blenders who achieve a specified reduction in carbon intensity of the blended fuel. It is thus a direct incentive to reduce carbon intensity in transport and other fuels. California's policy favors, for example, sugar-based ethanol and second-generation biofuels from cellulosic sources, such as switchgrass and *miscanthus*, over the maize-based ethanol currently favored under existing national policy.[12]

If a policy similar to the Low Carbon Fuel Standard—aimed at reducing carbon intensity by 15 percent—were adopted nationally and used to complement the existing Renewable Fuel Standard, simulations show that the amount of maize-based ethanol produced and consumed in the United States would fall by 11.8 billion liters by 2035, while ethanol from cellulosic feedstocks would increase by 12.5 billion liters by the same year.[13] Such a shift could have important implications for international markets and land-use change outside the United States.[14]

At present, a number of initiatives and studies are being conducted in the United States to see if it is feasible to scale up a California-like policy on a wider regional basis. A national low-carbon fuel policy would need to take into account the different fuel demands of the various subregions of the country in order to come up with a standard that could both lead to reduced use of high-carbon fuels and meet the concerns about energy security and affordability that are major components in the US debate about energy policy.

ROUNDTABLE FOR SUSTAINABLE BIOFUELS

Within the wider international community, there have been efforts to promote the sustainable production of biofuels and to provide producers with guidelines and incentives for ensuring that biofuels are low in carbon content relative to fossil-based alternatives, as well as compatible with international standards of decent work and fair compensation. Following the example of other initiatives for sustainable production, the Roundtable for Sustainable Biofuels was launched in 2011 as a mechanism for certifying biofuel producers who adhere to standards of low environmental impact and fair labor practices. This certification allows them to receive a price premium, similar to the price premium earned by fair-trade coffee producers. The Roundtable was designed in a way that makes the standards needed to reach official "sustainable" status compatible with those applied to certify the biofuels imported into Europe under the Renewable Energy Directive. Over time, the Roundtable for Sustainable Biofuels standards might be ratcheted up to encourage biofuel producers to further reduce the carbon intensity of biofuels; they could also include the indirect environmental effects of biofuels, especially those related to changes in land use and land cover.

LOOKING AHEAD

Food and energy markets will continue to interact in the future, creating fast-changing market opportunities for producers of feedstock crops like sugar and maize, regardless of whether they are supplying food, feed, or fuel sectors. But domestic trade policies and restrictions can lead to market disruptions and sharp price spikes, as seen in 2008 and 2010. One of the main lessons of the food price crises of the past several years is that open trade is essential to allow goods to freely flow to where their use is most highly valued and to allow for adjustments that will minimize market disruptions.[15]

In the European Union and the United States, the use of maize and rapeseed oil in biofuel production grew quickly until 2008/09, and this growth will likely stabilize as the US mandates for maize-based ethanol are met in 2015 and as policies in the industrialized countries encourage the use of alternative fuels with lower environmental impacts (see Figure 1). The significant rise in the use of maize in the United States—which nearly tripled over the period 2000 to 2009—implies that the maize market will be tight in the face of future changes, unless supply expands and grain inventory levels are rebuilt. Even though the rate of growth in rapeseed oil use is much smaller in comparison, its effect in tightening market conditions for vegetable oils will be similar.

The US Volumetric Ethanol Excise Tax Credit was scheduled to expire at the beginning of 2012, leaving the production and blending of ethanol to be driven by market forces. Given rising oil prices, it is likely that demand will continue pushing ethanol production above existing mandate levels. It remains to be seen whether the expiration of the tax credit will have an appreciable impact on US domestic feed prices for maize or on exports of US maize to the world market (which have, in turn, their own effect on world prices). As long as oil prices remain stable or rise slightly, there is little expectation that the profitability (and volume) of US ethanol production will be affected much. If demand for ethanol from Brazil or other countries rises, that would introduce an additional driver for ethanol production that is independent of the effects of any US policy instrument and would help remove the topic of ethanol subsidies from US political debate during this pivotal election year.

If the food price increases seen in 2011 persist into 2012, they will continue to provide increasing revenue to US grain producers while raising the cost of feed for livestock producers and of biofuel feedstock for US ethanol producers. Any US policies in 2012 to create incentives for the use of second-generation biofuel feedstocks (such as switchgrass, *miscanthus*, or other dedicated

FIGURE 1 US and EU use of maize and rapeseed oil in biofuel production and other industrial uses, 2000/01–2024/25

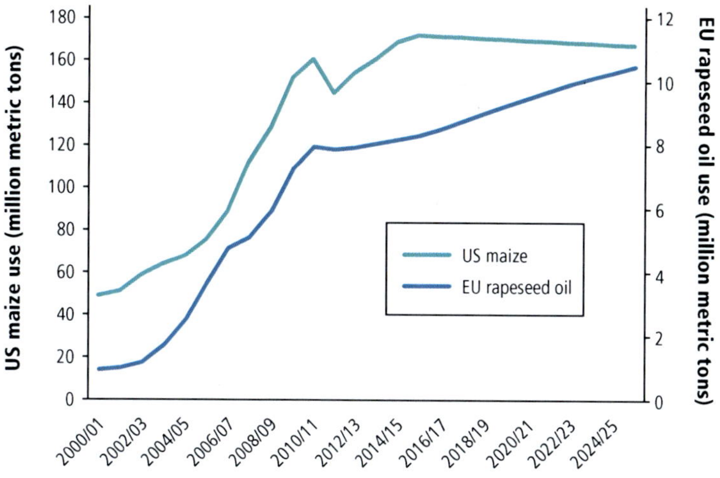

Source: Food and Agricultural Policy Research Institute, *FAPRI-ISU World Agricultural Outlook 2011* (Ames, IA: Iowa State University, 2011). **Note:** Maize use includes food, and rapeseed-oil use includes other industrial uses.

energy biomass) will promote innovation and encourage the US biofuels sector to diversify its sources of feedstock beyond grains to include agricultural residues (such as the maize stover left behind after harvesting). This may serve to relieve pressures on market demand for grain and on the land required to produce it. Careful assessment, however, is still necessary to measure the agronomic consequences of removing these residues from the field.

It is expected that the European Commission will make a formal biofuel policy recommendation in 2012 followed by a legislative process involving the European Parliament. Any decision by the European Commission will have global consequences because, besides changing the level and nature of biofuel production in Europe, it will serve as a model for lobbies and policymakers in many other countries.

Given the trajectory of the biofuels debate during 2011, policy discussions over the production and blending of biofuels seem likely to continue and will be fed by new research findings on the implications of biofuel policies for food security and the environment. ■

Chapter 6

AGRICULTURE, NUTRITION, AND HEALTH

Connecting the Dots

Rajul Pandya-Lorch, Heidi Fritschel, Zhenya Karelina, and Sivan Yosef, IFPRI

THE AGRICULTURE, NUTRITION, AND HEALTH nexus came to prominence in 2011. With 1 billion people continuing to suffer from food insecurity, and with vitamin and mineral deficiencies compromising the nutrition and health of billions of people, the international development community began to ask how much more could agriculture do to improve human well-being if it explicitly included nutrition and health goals? What kind of changes could maximize agriculture's contribution to human health and nutrition, and how could improved human health and nutrition contribute to a more productive and sustainable agricultural system?[1]

Although the agriculture, health, and nutrition sectors all seek to improve human well-being, agriculture has rarely been explicitly deployed as a tool to address nutrition and health challenges. With agriculture moving higher on the global agenda, in part because of volatile food prices, there is growing recognition that it is an opportune time to bring together the agriculture, nutrition, and health sectors and unleash the potential of agriculture—as a supplier of food, a source of income, and an engine for growth—to sustainably reduce malnutrition and ill-health for the world's most vulnerable people (see Box 6).

CREATING MOMENTUM AND BUILDING ON IT

Early in the year, about 1,000 leaders and practitioners in the sectors related to agriculture, nutrition, and health came together at an international conference called "Leveraging Agriculture for Improving Nutrition and Health," organized by IFPRI and its 2020 Vision Initiative in New Delhi (http://2020conference.ifpri.info/). At this conference, participants took stock of available knowledge on the interactions among agriculture, nutrition, and health; explored opportunities for enhancing nutrition and cutting health risks along the value chain; identified key levers and incentives for leveraging agriculture; and assessed critical research and action gaps. Ultimately, they catalyzed a process to reimagine how to make these linkages work better to enable more nutrition- and health-friendly agricultural investments (see Box 7).

BOX 6

Agricultural Research Takes on the Nutrition and Health Challenge

John McDermott, CGIAR Research Program on Agriculture for Improved Nutrition and Health

Malnutrition and disease are widespread and persistent global challenges. Agriculture is central to both, but agricultural growth alone has been insufficient to achieve targets for reducing malnutrition and improving health, such as United Nations Millennium Development Goal 1 on underweight children or Millennium Development Goal 4 on child mortality. One-third of children in South Asia are underweight, and more than 33 percent of childhood deaths in low-income countries are linked to undernutrition, most significantly in rural Sub-Saharan Africa. To enhance the agricultural contribution, the Consultative Group on International Agricultural Research (CGIAR) has developed a program to research agricultural actions for improving human nutrition and health.[1]

This new research program, launched in January 2012, has four interlinked components. One integrates agriculture, nutrition, and health programs and policies, while the other three components focus specifically on developing agricultural solutions that improve nutrition and health:

- Production and distribution of more nutritious staple crops, biofortified with pro-vitamin A, iron, or zinc, to address the most severe micronutrient deficiencies
- Improvement of value chains to increase foods' nutritional value from production to consumption, including food-value-chain analysis and development done by other CGIAR programs
- Reduction of the risk of agriculture-associated diseases by enhancing food safety and controlling zoonoses as well as emerging diseases, and by mitigating diseases associated with agricultural intensification

Research outputs will contribute to development impacts along three pathways: improving the nutritional quality and food safety of food value chains, providing knowledge and technologies to improve the performance of agriculture-nutrition-health development programs, and providing knowledge and evidence for improved policymaking and investment decisions.

For better nutrition and health for the poor, agricultural researchers will need to work closely with nutrition and public health researchers and link with food-value-chain actors, development program implementers, and policymakers. Behind these partnerships will be a fundamentally new perspective on agri-food system research and development, including

- looking beyond food production to processing, distribution, and consumption through deeper engagement with the private sector and other value-chain actors;
- taking a more integrative view through joint efforts of agriculture, health, and social development sectors using new metrics and tools for joint planning and assessment; and
- focusing on the perspective of the poor—by, for example, assessing livelihood and risk tradeoffs rather than using the standard hazard-avoidance perspective.

This new agricultural research program will focus on South Asia and Sub-Saharan Africa. Through investing in new tools, approaches, and evidence to usefully guide agricultural policy and practice, the CGIAR expects to have a major impact on enhancing agricultural contributions to global, regional, and national efforts to accelerate better nutrition and reduce agriculture-associated disease burdens among the poor.

Several development agencies have begun to design or redesign their programs to better tap these links. For instance, Feed the Future, the United States' multibillion-dollar global hunger and food security initiative, explicitly seeks to accelerate inclusive agriculture sector growth and improve nutritional status through sustainable country-owned development programs. The United Kingdom Department for International Development has substantially scaled up its support for nutrition programming and research and is including agriculture, food, and nutrition security research as part of its program in South Asia.

In Sub-Saharan Africa, the New Partnership for Africa's Development and the Global Alliance for Improved Nutrition signed an agreement to develop a five-year joint program to fully integrate nutrition security into the Comprehensive Africa Agriculture Development Program (CAADP) framework. Representatives from the ministries of agriculture, nutrition, and health and other counterparts from 17 West African countries came together at a CAADP workshop in Dakar in November 2011 to examine how nutrition can be integrated into national agricultural development plans, with special attention to addressing country-specific nutrition problems. In October 2011 President Yoweri Museveni launched the Uganda Nutrition Action Plan (2011–2016), developed by the Uganda National Planning Authority in collaboration with several ministries, with a strong message to the public on what foods to grow to avoid malnutrition. Malawi organized a groundbreaking national conference in September 2011 that brought together policymakers and planners in the agriculture, nutrition, and health sectors to coordinate and integrate their activities to help agriculture in Malawi contribute to the health and nutrition of the population.

In late 2010 a road map was produced for the Scaling Up Nutrition (SUN) movement—a broad partnership of international and donor organizations.[2] The movement gathered considerable momentum during 2011 when the road map began to be translated into action. By January 2012, 24 high-burden countries had committed to the SUN movement and begun setting nutrition goals and targets. More than 100 organizations around the world have endorsed

BOX 7

IFPRI's 2020 Conference: Tracking the Outcomes

Robert Paarlberg, Wellesley College and Harvard University

The 2011 “”Leveraging Agriculture for Improving Nutrition and Health” Conference, organized by IFPRI and its 2020 Vision Initiative, had significant useful effects on participants, in addition to informing global discourse and potential new initiatives. Conferees learned how to advance an integrated approach to agriculture, nutrition, and health more effectively in their respective workplaces. While most conferees arrived already believing the sectors should be viewed and managed jointly rather than in isolation, their attendance strengthened these opinions—as shown by pre- and post-conference surveys. Conferees gained valuable new information and connected to a wider set of cross-sector networks.

The 2020 Conference also produced measurable impacts on public and professional discourse. Between October 2010 and May 2011, the international journalists invited to the conference wrote 33 stories about the conference, and 25 other media stories were published in English, French, and German. Significant institutional reporting on the conference included 22 stories presented in various donor and stakeholder outlets. This media coverage helped increase the visibility of conference themes. Google searches at regular intervals revealed a significant uptick in the Internet presence of the conference's central theme; the average number of retrieved web pages containing the phrase “linking agriculture, nutrition, and health” increased from about 9,300 in the preconference period to more than 13,500 in the post-conference period.

Finally, surveys and interviews revealed that this New Delhi conference inspired or supported a range of important initiatives, including follow-on meetings and consultation; efforts to contact government decisionmakers on agriculture, nutrition, and health issues; new initiatives by donors; and even some provisional programmatic and institutional change. One immediate, tangible impact was a decision by the Canadian International Development Agency to give an additional US$6–10 million grant to the HarvestPlus project on biofortification. In addition the conference further strengthened the agriculture, nutrition, and health themes in the new CGIAR Research Program on Agriculture for Improved Nutrition and Health, an international initiative to create a network of educational institutions working in the areas of agriculture, nutrition, and health. China's State Food and Nutrition Consultation Committee vowed to create a food safety and nutrition development institute as well.

The durability and extent of such changes during the longer term will depend in part on whether IFPRI commits resources to sustained leadership in the areas of agriculture, nutrition, and health outreach and policy research.[1]

it. The movement supports national governments in developing and operationalizing nutrition-sensitive national plans and aligns financial and technical support for nutrition. A large part of the SUN movement's approach consists of incorporating specific pro-nutrition actions into other areas such as food security, agriculture, and health.

Other initiatives included the United Nations high-level meeting on noncommunicable diseases in September 2011. This meeting involved only limited participation by the agriculture sector, but the declaration that resulted from the meeting noted the need for a whole-of-government approach that includes the agriculture sector. With its report *Bringing Agriculture to the Table: How Agriculture and Food Can Play a Role in Preventing Chronic Disease,* the Chicago Council on Global Affairs provided clear analysis and recommendations on how agriculture can contribute to better health.

Despite opportunities to improve health outcomes through the agriculture nexus approach, involving the health sector in the discussions has been challenging (see Box 8). One of the key barriers to collaboration between the agriculture and health communities is a lack of common metrics. Therefore, in May 2011, IFPRI and the Leverhulme Center for Integrative Research on Agriculture and Health brought together health and agriculture experts to find common ways of measuring the health outcomes of agriculture interventions.

Building on the momentum of the 2020 Conference, the Consultative Group on International Agricultural Research (CGIAR) developed a major new research program called "Agriculture for Improved Nutrition and Health," which was launched in January 2012 with the overarching aim of improving the nutrition and health of poor people by exploiting the many synergies between agriculture, nutrition, and health (see Box 6).

LINKING AGRICULTURE, NUTRITION, AND HEALTH

In many ways, the links among agriculture, nutrition, and health are already at work, but the synergies may not always be optimal. Agriculture is the primary source of food to meet people's need for energy and essential nutrients. But to get access to food, people do not necessarily need to produce it themselves; they can also buy it. The agricultural system may help increase people's access to food by allowing them to produce more food (if they farm themselves) or by lowering food prices or raising their incomes (if they purchase food). By improving their access to food, agriculture has the potential to greatly improve people's nutrition and health. At the same time, some agricultural conditions and practices can lead to disease and poor health for both farmers and consumers.[3] For example, agricultural practices may increase farmers' risk of becoming infected with animal diseases, expose farmers to dangerous pesticides, or introduce toxins into foods.

In many agrarian countries, agricultural growth is more effective in reducing undernutrition than growth in other sectors. However, the composition of agricultural growth, the distribution of this growth, and the conditions under which such growth takes place all matter. Growth in agricultural subsectors where poor people are engaged, such as staple crops, contributes more to reducing poverty and increasing calorie intake than growth in, for instance, export crops. Later in the development process, growth in other sectors besides agriculture becomes more important in improving food and nutrition security. Yet neither agricultural growth nor nonagricultural growth alone is sufficient to reduce child undernutrition or micronutrient malnutrition—complementary programs in nutrition, health, water and sanitation, and behavior change communication also need to be implemented and targeted to vulnerable populations, especially women and children.[4]

The links among agriculture, health, and nutrition often work differently for men and women. In many parts of the world, men and women spend money differently: women are more likely to spend the income they control on food, healthcare, and education for their children. Increased equality between men and women can translate into greater agricultural productivity. If this productivity is accompanied by more income and strong bargaining power for women, it can result in better health and nutrition.

Opportunities to improve nutrition and reduce health risks exist all along the agricultural value

BOX 8

Public Health and Agriculture: Working Together

Kabba T. Joiner, Helen Keller International

The agriculture and health sectors have long been separated by fundamentally different societal functions and institutional organization. However, both health and agriculture representatives made a marked effort to bring the two sectors closer together in 2011, forming some promising links between agriculture and health organizations. Programs that have emerged more recently in Sub-Saharan Africa include the Baby Friendly Community Initiative in The Gambia, Millennium Villages in Mali, Gardens for Health in Rwanda, and Agriculture for Children's Empowerment in Liberia.

Agriculture can make both direct and indirect contributions to health. Growth in agriculture leads to increased rural income, which is positively related to better health status when community health infrastructure is financed by profits from agriculture. Sustained agricultural development can indirectly lead to significant progress in rural health. In particular, if women's incomes grow, they use healthcare services more frequently, which improves maternal and child health.

Agriculture can contribute to public health directly through improved agricultural products. In general, improving diets—by improving food products—reduces the burden of chronic diseases. Integrating the agriculture and health sectors also improves food safety by making it possible to establish better surveillance systems from farm to table. But more can be done than just growing better-quality crops. For a long time, agriculture was not considered a primary weapon in the elimination of micronutrient malnutrition. Food systems were developed with little attention to balanced nutrient requirements that support good health and well-being. Now HarvestPlus and other organizations are addressing this issue through the breeding of mineral- and vitamin-rich crops, such as orange-fleshed (that is, carotene-rich) sweet potatoes and high-iron pearl millet.

Collaborations between the agriculture and health sectors can lead to substantial improvements in diet quality in developing countries, but they can flourish only if certain human and institutional challenges are overcome. Representatives from both sectors need to

- take cross-sectoral action at the community level;
- increase funding in units of the health sector that can work with agriculture;
- create formal arrangements, assign responsibilities, and develop skills for intersectoral negotiation and decisionmaking;
- establish reliable communication and links among researchers, policymakers, and practitioners in both sectors;
- ensure mutual consultation in priority setting and activities like data collection; and
- strengthen human capital in both sectors by reviewing curricula or by exchanging staff and sharing facilities.

Decisionmakers in agriculture and health should push for more innovation and cross-sectoral participation to produce better outcomes. They must go off the beaten path in order to maximize the benefits from their collaboration.

chain. A value-chain approach to development can incorporate nutrition goals and thereby make nutritious foods more available and affordable for the poor. This approach starts by looking at every component of the food supply chain from field to fork—including production, postharvest processing, marketing, and trade—and determining where value for nutrition can be integrated. The food value chain also involves many hazards—microbiological, physical, and chemical hazards, as well as occupational hazards—that pose challenges for producing and consuming safe food. Policymakers are increasingly using risk analysis to help them decide on regulatory and other actions to reduce health risks along the food value chain.[5]

Many interventions are being tried to understand and deal with these challenges. Examples include biofortification (the breeding of new varieties of food crops with improved nutritional content); schemes to increase household production and consumption of micronutrient-rich vegetables, fruits, and animal-source foods; local production of foods for school feeding programs; and projects to integrate agriculture, nutrition, and health services.[6] So far, however, there is little concrete evidence on how agriculture–nutrition linkages work. One crucial task then is to compile the evidence base on these links. Many more studies are needed on the nutritional impacts of agricultural interventions, more nutrition-relevant data need to be generated and collected, and nutritional indicators should be included in evaluations of agricultural programs.

The 2020 Conference highlighted four important sets of tools that could help to leverage agriculture for better nutrition and health. Economic levers include, at the broadest level, agricultural growth or overall economic growth (with the caveat that growth alone is not enough to solve the nutrition problem). "Fat taxes" and "thin subsidies" have the potential to influence people's economic access to healthy foods in industrial counties, but more targeted approaches to improving poor people's diets may be more appropriate in developing countries. Social levers involve bringing people together across sectors and within communities to jointly work toward improving nutrition and health. Governance levers require government leadership at all levels—from national to provincial to local. Changes in policies and programs are not enough to get people in different ministries and institutions to work together—it is important to devise incentives to get them to do so and to devote the time and resources necessary to work across sectors. Science and technology levers require not only allocating more resources to general agricultural research and development to keep the pipeline for innovation, discovery, and dissemination full, but also targeting more resources specifically to nutrition- and health-relevant research, such as work on nutrient-rich vegetables and other crops and livestock.

A number of recurring themes[7] emerged during the 2020 Conference and are engaging the international community:

1. *Improve investments* by making existing ones more nutrition- and health-friendly, prioritizing and scaling up successes, and generating new ones that exploit the links among agriculture, nutrition, and health.
2. *Don't wait to act* but move ahead based on available information and common sense.
3. *Communicate better* to build awareness, raise interest, provide options, and attract "champions" to promote action.
4. *Fill the knowledge gaps* on what type of agricultural growth is best for nutrition and health and what types of governance arrangements and partnerships are needed at the local, regional, and global levels.
5. *Focus on education* by developing multidisciplinary university-level education programs that inculcate broader thinking among future leaders in agriculture, nutrition, and health, and break down the "silos" between the sectors.
6. *Build the evidence base* by collecting relevant data in a timely fashion, improving tools and methods, and investing in monitoring and evaluation.
7. *Collaborate across sectors* by creating mutual accountability and looking for ways to work together while not losing the advantages of deep sectoral expertise.
8. *Use all available levers for change,* including economic, social, governance, and science and technology levers that can maximize agriculture's contribution to nutrition and health.
9. *Correct market failures* by using public policies such as investments, subsidies, education, trade, and tax policies, as markets alone may not achieve socially optimal agriculture, nutrition, and health outcomes.
10. *Look at food systems, not just agricultural systems;* consider all the stages from field to fork; and be sensitive to the sustainability of natural resources.
11. *Proactively engage the health sector* and find ways to reach out and include the health sector in agricultural activities.
12. *Recognize that women are at the nexus of the three sectors* and direct policies and programs to women to simultaneously strengthen agriculture and enhance nutrition and health.

The nexus approach is spilling over to other sectors. The food-water-energy nexus gained a great deal of attention in late 2011 with the Bonn2011 Nexus Conference (see Box 9). In an increasingly interlinked global environment, a nexus approach to agriculture offers considerable potential to improve nutrition and health, to manage natural resources more sustainably, to improve people's livelihoods, and to support more inclusive economic growth. Looking ahead, it is important to build an evidence base that will improve understanding and help identify viable opportunities to strengthen linkages across sectors and achieve mutually beneficial outcomes. ■

BOX 9

Food, Water, and Energy: Understanding the Nexus

Claudia Ringler, IFPRI

During the last few years, the cross-sectoral linkages on the supply side of agriculture have become more apparent as key agricultural inputs have grown scarcer and more expensive. Key among these linkages are those of agriculture and food with water, land and energy resources, and environmental/biodiversity outcomes. The food-water-energy nexus has come to the forefront in discussions at several international forums in the run-up to the Rio+20 United Nations Conference on Sustainable Development that will take place in Brazil in June of 2012. One such forum, the Bonn2011 conference on "The Water, Energy, and Food Security Nexus: Solutions for the Green Economy," concluded that "achieving water, energy and food security, and consequently reducing hunger and eradicating poverty, is a central future challenge that is possible to overcome, even under difficult and challenging global economic conditions."[1]

Much work has been done on water and food interlinkages. Water supply is essential for food production, which depletes about 80 percent of global freshwater withdrawals annually. Population growth, economic growth, urbanization, and industrialization have fueled increasing water scarcity, putting as much as half of all global grain production at risk of insufficient water resources by 2050.[2] Increasingly it is not only water availability that is being compromised, but also water quality. Investments in the sector have been insufficient in most developing countries to meet growing demand for clean and safe water.

Less is known about the interlinkages between energy and food and among energy, water, and food. However, the growing interdependence of food and oil prices as a result of increased energy use in agriculture and the growing share of foodcrop use as biofuels have made the need for joint policy development apparent. Higher energy prices have driven up food prices and reduced the availability of land and water for food production (due to competition from expanded biofuel production). At the same time, poor people's access to sufficient food, water, and energy remains unacceptably low, particularly in Sub-Saharan Africa and South Asia.

These linkages thus demand holistically developed programs and policies. This is particularly crucial because food production will need to increase substantially in the next four decades to meet growing demand. To achieve food security without compromising sustainable water and energy supplies, improved policies, institutions, and investments should include the following principles:

- develop clear national food and nutrition policies that take into account the consequences for water and energy;
- reduce water, food, and energy subsidies that lower resource-use efficiency and have adverse impacts on the poor and the environment;
- maximize complementarities between public and private stakeholders in food, water, and energy provision;
- promote resource-use-efficient technology development and dissemination, particularly technologies the poor can afford;
- promote tenure security for both water and land;
- focus and strengthen crop and other agricultural research at the food-water-energy nexus (for example, drought-tolerant, high-yielding, nutrient-use-efficient crops); and
- create markets and trade solutions that ensure least-cost input flow for farmers and consumers.

If food, water, and energy connections remain unaddressed, global food security will not be achieved, particularly for the rural poor.

Chapter 7

LAND DEGRADATION

Land under Pressure

Ephraim Nkonya, Jawoo Koo, and Paswel Marenya, IFPRI;
Rachel Licker, University of Wisconsin, Madison

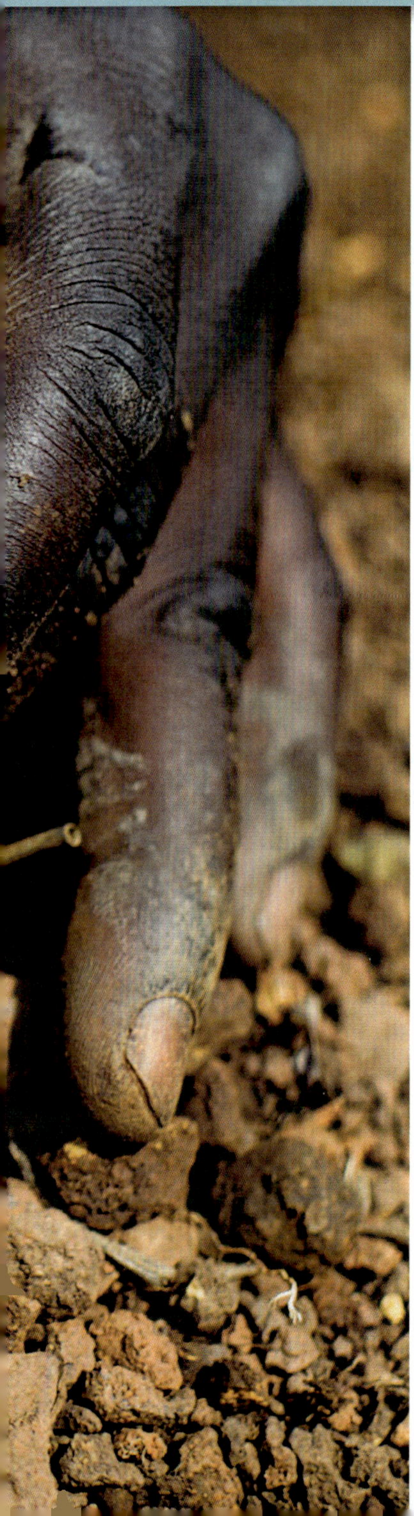

IN OCTOBER 2011, THE GLOBAL POPULATION reached 7 billion people, a milestone that highlights the enormous pressure on the planet's ecosystems. In the face of this population growth, the farmland on which global food production depends is degrading rapidly. About 24 percent of global land area has been affected by land degradation. This area is equivalent to the annual loss of about 1 percent of global land area, which could produce 20 million tons of grain each year, or 1 percent of global annual grain production. Globally, 1.5 billion people and 42 percent of the very poor live on degraded lands.[1]

Population is growing fastest in the developing countries. In Sub-Saharan Africa, annual population growth of 2.3 percent has led to ever smaller areas of arable land per capita. Between 1961 and 2009, per capita arable land in Sub-Saharan Africa fell by about 76 square meters a year, the steepest drop in the world (Figure 1).

How can this crowded world feed growing numbers of people? The most feasible solution is to sustainably raise agricultural productivity on existing land. This means halting land degradation to avoid losing even more valuable farmland. It also means raising soil fertility on existing farmland to boost yields and addressing other challenges, on and off farms, that have contributed to low agricultural productivity.

ACHIEVING ZERO LAND DEGRADATION

In September 2011 the United Nations General Assembly called for building a world with no land degradation. And, in October 2011, parliamentarians of the United Nations Convention to Combat Desertification reaffirmed this goal by

FIGURE 1 Annual loss of per capita arable land in developing countries, 1961–2009

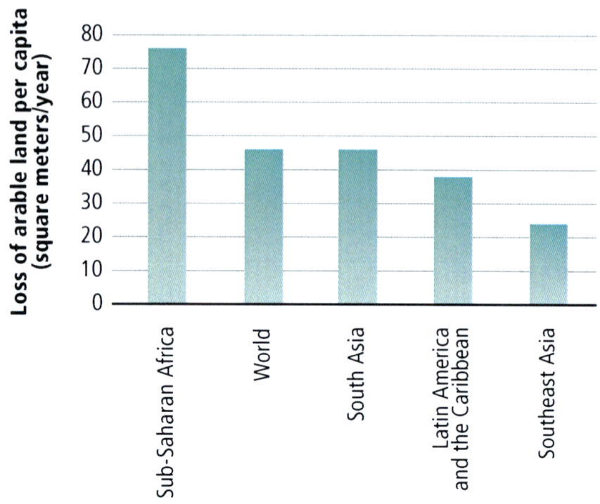

Source: Preliminary analysis based on linear regression model from data from Food and Agriculture Organization of the United Nations, FAOSTAT database.

issuing a declaration calling for zero land degradation and for adopting sustainable land management as the way to achieve sustainable development.

New Evidence

In pursuing these goals, we can draw on new evidence about causes of land degradation and solutions to it. Although the conventional wisdom holds that increasing population density is a major driver of land degradation, recent studies suggest that land improvement is possible even in countries with high population density. For example, there was a positive association between an increase in population density and land improvement between 1981 and 2006 globally and in East Asia, but a negative association in Sub-Saharan Africa and South Asia.[2] The land improvement in East Asia was attributed largely to policies promoting tree planting and forest plantation programs in China and Korea. China, for example, has the largest number of certified emission reduction credits in the world. These credits, issued to developing countries that implement carbon-reducing projects, are certified according to standards set by the Clean Development Mechanism of the Kyoto Protocol.

More is also being learned about how effective governance—especially local governance—can reduce land degradation. Strong local institutions and regulation, linked with national institutions and policies, can give local communities the tools they need to manage natural resources such as land, water, and forests. For example, a 2011 study of four African countries showed that the number of land and water management by-laws enacted by communities was strongly correlated with the level of decentralization in each country.[3] The findings suggested that the more decentralized a country is, the more likely local communities will take collective action to address natural resource challenges. Of course, strong local governments and other local organizations alone are not sufficient for sustainable natural resource management. Households and communities also need access to markets, remunerative prices, and other incentives.

In the past 20 years, international cooperation on natural resource management has risen rapidly because of increasing awareness of how the benefits of better resource management and the costs of resource degradation cross boundaries. Carbon markets—valued at US$142 billion in 2010[4]—and other international environmental programs have contributed to global efforts to reduce deforestation and other sustainable natural resource management programs in developing countries.

The Example of Brazil

Brazil has dramatically reduced deforestation, thanks to actions taken by local and central governments in cooperation with the international community. In the 1990s and early 2000s, Brazil enacted laws and regulations to protect forests, but deforestation was rampant, reaching a peak of 72,000 square kilometers in 2003–04.[5] Then state and municipal governments, as well as the federal government, set laws and by-laws enforcing the deforestation moratorium. And Brazilians started participating in Clean Development Mechanism projects. With more than 180 such projects in operation, Brazil has the third-largest number of certified emission reduction credits, after China and India.[6] Land users who signed a binding contract promising not to clear forests were compensated. By 2008–09, just five years later, deforestation had plummeted to 7,000 square kilometers—a reduction of 74 percent.[7] Moreover,

BOX 10

India: Common Lands and Food Security

Jagdeesh Rao Puppala and Rahul Chaturvedi, Foundation for Ecological Security

In a landmark judgment issued in January 2011, India's Supreme Court ruled that arrangements for effective self-governance of community property at the local level be made and that those encroaching on "commons" be evicted. "Commons" refers not only to common pool resources like forests, pastures, riverbeds, bodies of water, and farmland owned jointly by the community and enjoyed by all but also to genetic material such as seeds and livestock breeds adapted by local communities. These shared resource systems and village governments are crucial to sustaining the farming systems—and, thereby, the livelihoods and food security—of the more than 300 million people living and working in rural communities in India, where common lands are estimated to constitute roughly 15 to 25 percent of the geographical area.

In response to the Supreme Court judgment, some state governments have already taken measures to improve tenure security, strengthen institutional arrangements for local governance, and restore common lands. At the national level, preliminary policy planning also calls for a commons policy and increased public investments to help build common property regimes.[1]

In addition to contributing to food and nutritional security, commons help maintain critical ecological functions for the sustenance of farming systems. Forest commons play a crucial role in maintaining stream flows, groundwater recharge, and nutrient transfers. They serve as habitats for pollinators and pest predators, thereby building the resilience of agro-ecosystems, and provide an institutional setting that energizes collective action while minimizing undesirable individual action that can lead to the exploitation of resources.

A 2010 study in rainfed areas of India underscores the continuing dependence of rural households on commons: across the 3,000 households surveyed, 53 percent accessed commons for agricultural inputs, 69 percent for livestock grazing, 62 percent for domestic and livestock water requirements, 74 percent for fuelwood collection, and 37 percent for food items for household consumption.[2] Roughly 45 percent of the total fodder requirement is met by common resources, and commons are estimated to contribute 20 to 40 percent to household annual incomes. Despite all of this, commons have long been neglected in policies and programs that aim to restore natural resources. However, recent developments, including the 2011 Supreme Court ruling, indicate that this is beginning to change, and the role of local communities in effectively governing natural resources is receiving recognition and leading to more decentralized governance.

while Brazil's agricultural area has remained relatively unchanged, at 68.5 million hectares, since 2005, food production in Brazil appears to have increased by 11–17 percentage points between 2007 and 2009.[8] This is a good example of how government policies and strategies can help agriculture-based economies achieve sustainable land management in collaboration with the international community.

SUSTAINABLY INCREASING AGRICULTURAL PRODUCTIVITY

Besides halting land degradation, it is crucial to produce more food on current farmland. Researchers have shown that land productivity will have to meet more than three-quarters of the growth in global food demand between now and 2050.[9] Meeting the increase in food demand will mean raising crop productivity in regions where there is a wide gap between actual yields and potential yields. In the developed regions and East Asia, growth in crop yields is slowing as the gap between potential and actual yields narrows. For the three major cereals—rice, wheat, and maize—the largest yield gaps occur in Sub-Saharan Africa, Eastern Europe and Central Asia, and South Asia.[10]

In Sub-Saharan Africa, which has the widest gaps between actual and potential yields, these gaps can be closed by investing in, for example, agricultural research, improvements in market conditions, and better rural services, which will provide technical support and incentives for increasing productivity. But among the most important steps to increase agricultural productivity and address land degradation is managing soil fertility.

Organic inputs such as manure and crop residues have been shown to raise crop yields and improve soil ecology.[11] By increasing soil's water-holding capacity and carbon content, organic inputs can help both adapt to and mitigate climate change. Organic soil fertility management can also reduce the amount of nitrogen fertilizer required. One study found that when leguminous trees were planted on maize plots in East and Southern Africa, the maize required up to 75 percent less synthetic nitrogen fertilizer.[12] Reduced use of nitrogen fertilizer could, in turn, reduce the water pollution that results when farmers apply too much fertilizer.

In fact, research conducted in Sub-Saharan Africa has shown that integrated soil fertility management—the use of organic inputs, improved crop varieties, and judicious amounts of synthetic fertilizers—is more profitable than the use of organic inputs or synthetic fertilizers alone. A recent study conducted in Kenya, Malawi, Mali, Nigeria, and Uganda showed that as farmers applied more organic inputs and fertilizers, both maize yields and soil carbon increased. Over a 30-year period, farmers who combined synthetic fertilizer and organic inputs experienced increasing or flat maize yields, whereas farmers who used organic inputs or synthetic fertilizer alone experienced falling yields. Further analysis showed that by using integrated soil fertility management, farmers could increase the financial return to applying one kilogram of nitrogen by an average of about 60 percent, compared with use of fertilizer alone.[13] Countries that allocate large shares of their agricultural budgets to fertilizer subsidies may be able to increase the returns to public spending by giving subsidies on the condition that the beneficiary adopts an easily verifiable organic input practice such as agroforestry.[14]

Despite its win-win-win advantage of greater yield, profit, and environmental services, adoption of integrated soil fertility management is low—often lower than the use of synthetic fertilizer or organic inputs alone (Table 1). There may be several reasons for this low adoption rate. Farmers are less likely to adopt this practice when they have little household capital, are far from roads and markets, and have no livestock to produce manure and help transport bulky inputs. According to preliminary research results, women farmers are more likely to use organic inputs and less likely to use synthetic fertilizer than men, but greater endowments of financial and physical assets increase their propensity to adopt integrated soil fertility management practices.

The capacity of agricultural extension services in Sub-Saharan Africa to provide integrated soil fertility management technologies is low. Only a small share of agricultural extension agents in Nigeria and Uganda, for example, were shown to provide advisory services on organic inputs. Most of their agricultural advisory services focused on improved seeds, fertilizers, and pesticides.[15]

CONCLUSION

In 2011 two United Nations bodies adopted ambitious goals for halting land degradation and achieving sustainable development. These

TABLE 1 Adoption of integrated soil fertility management in six Sub-Saharan African countries

Soil inputs	Farmers' rate of adoption (%)					
	KENYA	MALAWI	MALI	NIGER	NIGERIA	UGANDA
Organic inputs and synthetic fertilizer	19.6	14.7	17.7	0.0	7.5	2.0
Organic inputs alone	29.4	18.6	38.7	1.0	12.1	11.9
Synthetic fertilizer alone	10.5	70.8	16.3	0.1	45.3	6.1

Sources: Kenya, Niger, Nigeria, and Uganda: Compiled from E. Nkonya, F. Place, J. Pender, M. Mwanjololo, A. Okhimamhe, E. Kato, S. Crespo, J. Ndjeunga, and S. Traore, *Climate Risk Management through Sustainable Land Management in Sub-Saharan Africa,* IFPRI Discussion Paper 01126 (Washington, DC: International Food Policy Research Institute, 2011); Mali: Government of Mali, *Recensement general de l'agriculture, Campagne agricole 2004–2005;* Malawi: Ministry of Agriculture and Food Security household survey 2008. **Note:** In this case, organic inputs consist of animal manure. Farmers are considered to be using integrated soil fertility management when they adopt both organic inputs and synthetic fertilizer (all farmers had already adopted improved crop varieties).

BOX 11

Women's Land Tenure Matters

Ruth Meinzen-Dick, Amber Peterman, and Agnes Quisumbing, IFPRI

Secure land tenure is widely recognized as crucial for investments, productivity, sustainability, and status. Without secure rights over their land, people do not have the incentive to invest in long-term sustainability or productivity enhancements on their land and may not even have the authority to make investments as straightforward as planting trees. Secure tenure reduces vulnerability to eviction and strengthens bargaining power, both outside and within the household. Although most analysis of land tenure has focused on the household, there is increasing acknowledgment that the gender distribution of land rights both across and within households also matters. The 2010–11 *State of Food and Agriculture Report* highlights how the gender gap in access to productive resources constrains agricultural productivity.[1] Other reports and organizations have also shown that control of land is important for women's security, status, and bargaining power.[2] The 2012 *World Development Report* recommends strengthening women's land rights as an important action for improving gender equality, a development objective in its own right that will also enhance productivity and development outcomes for the next generation.[3]

Currently, women are less likely to be landholders than men. When women do have access to land, they usually cultivate smaller and less fertile holdings. According to the Gender and Land Rights database maintained by the Food and Agriculture Organization of the United Nations, on average women account for less than 5 percent of agricultural landholders in West Asia and North Africa, 15 percent in Sub-Saharan Africa, and 25 percent in Latin America.[4] But even this is an incomplete picture of legal status because landholding may not confer many rights to the female landholder. A recent study in Uganda suggests that, even when men and women reported holding land jointly, women were much less likely to have any documentation in their name.[5] This would make them susceptible to losing land rights if their husband decided to sell the land or if they were divorced or widowed.

What can be done to strengthen women's land rights? In many developing countries, legal reforms are often ignored if they run counter to customary law and practice. Therefore, reforming the legal system is important but unlikely to have much effect on its own. The implementation of reforms such as removing restrictions on women's land ownership, making provisions for joint titling, and reforming family law so that women can inherit land and retain rights in case of divorce or widowhood requires attention and resources. This may involve offering legal literacy programs to inform administrators and judges, as well as the general public, about the reforms and working with customary authorities to discuss the importance of women's land rights. Including women on local land administration committees can encourage them to register their land. An analysis in its early stages suggests that in Ethiopia's land registration process, having more women on the local land committees increased attendance at meetings relating to land registration and raised awareness of the new family code that aims to strengthen women's property rights.[6] This and other studies show that improving women's land rights should be done through multiple channels, rather than through simple, one-step solutions.

goals will be difficult, but not impossible, to meet. The evidence presented here suggests several avenues for achieving a world with no land degradation.

First, efforts to promote sustainable land management need to improve local and national governance while also enhancing international cooperation.

Second, instead of focusing solely on fertilizer subsidies, countries should use broader and more cost-effective incentives to encourage farmers to adopt integrated soil fertility management. ■

NEW PLAYERS

Stepping into the Global Food System

Kevin Chen and P. K. Joshi, IFPRI

Ensuring global food security has traditionally been perceived as the responsibility of a few industrial countries and United Nations agencies, but the strengthening of many developing countries and institutions is also contributing to the global agenda of ensuring food security, alleviating poverty and ending hunger. The emergence of new players has been a growing trend for several years. In 2011, key developments marked an unprecedented degree of influence by new players, including countries such as Brazil, China, and India; new institutions such as charitable foundations; and the private sector, on the global food governance system.

RISE OF THE EMERGING ECONOMIES AS NONTRADITIONAL DONORS

The emerging economies—particularly Brazil, China, and India—have grown at remarkably high rates in the past decade,[1] and in 2011 these three economies accounted for more than 20 percent of global gross domestic product (Figure 1). These are not the only emerging economies assuming roles as major global players. The Group of 20 (G20) countries, representing two-thirds of the world's population, 90 percent of world gross domestic product, and 80 percent of world trade, are quickly overtaking the G7 and the G8 as

FIGURE 1 **Emerging economies' share in global gross domestic product (GDP), population, and crop production, 2011**

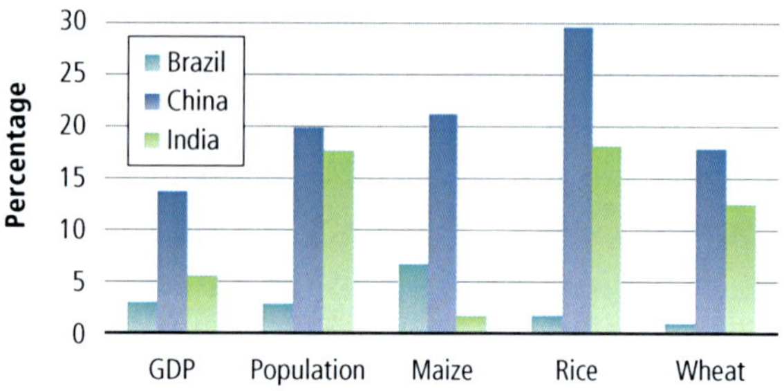

Source: GDP and population data are from the International Finance Corporation. Crop production data are from the Food and Agriculture Organization of the United Nations, FAOSTAT database, faostat.fao.org. **Note:** Share of crop production is for 2010.

the principal forum for managing global economic problems. These emerging economies are changing the structure and nature of the global landscape and global governance, and their important role in addressing global food security, in particular, was frequently acknowledged and discussed at high-level ministerial meetings in 2011.

On June 22 and 23, 2011, for example, the G20 agricultural ministers met in Paris to develop an action plan on food price volatility and agriculture. And on October 30, 2011, the agriculture ministers of Brazil, China, India, and Russia met in Chengdu, China, to discuss agricultural development and cooperation among themselves and with other developing countries. These ministers believe that the stable and robust agricultural development of their countries is important to world food security and see this as an especially critical strategy for reducing hunger in the South.[2] Emerging economies increasingly affect growth and development prospects in developing countries through direct links, such as aid, trade, and foreign direct investments, and through indirect linkages, such as commodity prices and competition in Third-World markets.[3]

In December 2011, the Fourth High-Level Forum on Aid Effectiveness, held in Busan, South Korea, highlighted the increasing importance of South–South cooperation in development.[4] South–South cooperation is now seen as a seamless part of the International Fund for Agricultural Development's country program.[5]

A number of emerging economies—including Brazil, China, India, Malaysia, South Africa, Turkey, and a number of countries in the Middle

BOX 12

China and African Agriculture: Rumors and Realities

Deborah Brautigam, IFPRI

Separating rumors from realities about Chinese engagement in African agriculture is not easy. Reports have claimed the Chinese set up a US$5 billion fund for African agriculture, sent a million farm workers to Africa, pledged US$800 million to modernize agriculture in Mozambique, or farmed huge tracts in Zimbabwe or Ethiopia. Fieldwork by several researchers presents a different picture: Chinese engagement in African agriculture is both more diverse and smaller than is often believed.

Take Chinese aid, for example. Between 1964 and 2009, Chinese aid teams constructed at least 142 agricultural projects, including state farms, irrigation schemes, and demonstration centers for African governments. Yet these projects' poor sustainability caused the Chinese to revise their approach and give Chinese companies a leadership role.

Thus, in an experiment launched in 2006, Chinese aid is financing 20 agrotechnology research, training, and demonstration centers in Africa. A Chinese research institute or agribusiness firm is building each one at a cost of US$6 to 9 million. The centers will specialize in activities chosen by the host country. For example, Ethiopia wants its center to demonstrate the complete value chain for horticulture exports. China will provide additional aid to help run each center

East—have recently emerged as active partners in technical and economic cooperation in developing regions, especially Africa. Official development assistance—or foreign aid—from the emerging donors rose from US$4.6 billion in 2005 to US$10.4 billion in 2009 (Figure 2). These donors contributed about 10 percent of global aid flows in 2008.[6]

China is fast becoming a major investor in Africa, although Chinese engagement in African agriculture is both more diverse and smaller than is generally perceived (see Box 12). In 2000, China moved to consolidate this cooperation by establishing the Forum on China–Africa Cooperation, which meets every three years. As part of this initiative, China has significantly boosted its aid budget in recent years, with a stronger emphasis on agricultural development. Overall aid from China to Africa is estimated to have almost quadrupled from US$684 million in 2001 to US$2,476 million in 2009.[7] At the 2010 United Nations High-Level Meeting on the Millennium Development Goals, China pledged to establish 30 demonstration centers for agricultural technologies in other developing countries, dispatch 3,000 agricultural experts and technicians to these countries, and invite

FIGURE 2 **Aid from emerging economies, 2005–09**

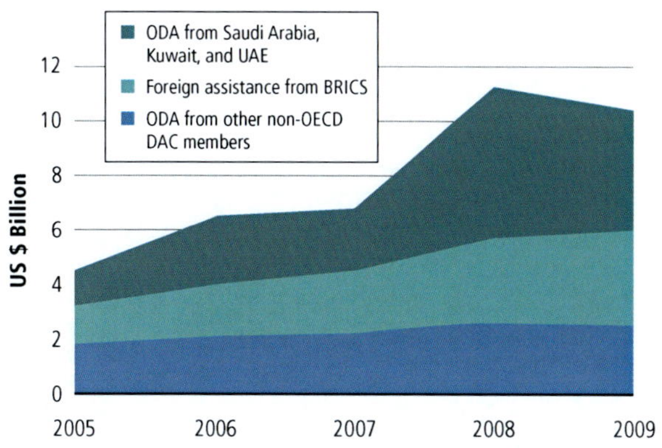

Source: Reproduced from K. Smith, *Non-DAC Donors and Humanitarian Aid: Shifting Structures, Changing Trends* (Somerset, UK: Development Initiatives, Global Humanitarian Assistance, 2011). **Notes:** Aid data are for official development assistance (ODA). OECD is Organisation for Economic Co-operation and Development; BRICS is Brazil, Russia, India, China, and South Africa; and DAC is Development Assistance Committee.

5,000 agricultural personnel from these countries to China for training. By 2011, China had already established 14 centers for agricultural research in a number of African countries.

India is also bolstering its cooperation with Africa. For example, the Africa–India Forum

for at least three years. During this time, Chinese experts will demonstrate how the centers can develop income-generation activities to boost sustainability (while also looking for new business opportunities for their institutes and firms).

In another experiment, Chinese companies began to lease some of the old Chinese aid projects in the 1990s, as they were privatized: Sukula sugar complex in Mali, Magbass in Sierra Leone, and Koba in Guinea, for instance. In 2004, China's Ministry of Commerce started to encourage country-specific opportunities for Chinese agricultural investment: cotton in Egypt, fruit and nuts in Nigeria, sisal in Tanzania, tobacco in Zimbabwe, and nonspecific crops in Zambia, Ethiopia, Guinea, Benin, the Democratic Republic of the Congo, and Cameroon.

As of yet, few of the existing Chinese investments in Africa appear to be larger than 5,000 hectares. Several larger Chinese biofuel projects proposed in Zambia (jatropha), Ethiopia (sugarcane), and the Democratic Republic of Congo (maize) have stalled or been abandoned. As a Chinese official commented in Tanzania: "Agriculture is risky. It is hard to have [a] 'win–win.'"

Land transfers frequently present food security risks for local communities, and large Chinese farms are no exception. However, surprisingly little evidence exists for the common assumption that the Chinese plan to use African land for China's own food security. China imports no grain from Africa—instead, cotton, sesame seeds, and tobacco head the list. Chinese agroprocessing companies have contractual partnerships with local smallholders who grow cotton (in Malawi, Mozambique, and Zambia) and tobacco (in Zimbabwe). The majority of Chinese farms appear to produce food for local markets. The evidence, at least for now, does not support the rumors.[1]

BOX 13

Brazil: An Emerging Power in Agriculture

Beatriz da Silveira Pinheiro and Geraldo B. Martha Jr., Embrapa

In 2011, Brazil's agricultural export surplus exceeded US$70 billion. The overall performance of Brazilian agriculture in the last four decades has transformed the country from a net importer of several products to one of the most relevant players in the international agricultural commodities market. To a great extent, the huge transformation was a response to an increased demand for agricultural products prompted by the industrialization process of the 1960s to the 1980s. This industrialization period was associated with a growing and increasingly richer and more urban population. The increased opportunity cost of labor for farmers led to a favorable environment for agricultural modernization. Huge investments were made in agricultural research and development efforts during the last four decades, even in the first years (1970s and 1980s) when the modernization of Brazilian agriculture was still just a promise.

As food production increased more than food demand, real food prices decreased: in the period 1975–2010, consumer food prices decreased by half. This huge drop in food prices, along with associated reduced price volatility, has alleviated inflationary pressures and ensured national food security. In addition, lower food prices have effectively boosted other sectors in the economy because paying less for food frees up more income, especially for the poor. Brazil strongly focused on technology-driven productivity gains, which became an important additional characteristic of agricultural production expansion. Without these gains, meeting 2006 production levels would have required an additional agricultural area 30 percent larger than the Amazon Biome in Brazil.

This vigorous increase in agricultural production also allowed for increased exports. In the last two decades, Brazil diversified exports, and oilseeds, grains, and meats acquired great relevance. The resulting export surplus has guaranteed positive results for the Brazilian balance of trade and supported food prices in domestic markets, with positive effects on Brazilian farmers' income. From a global perspective, the increased volume of Brazilian agricultural exports has made an important contribution to reducing world hunger and food-price inflationary pressures in developing countries.

Brazil's experience in producing agricultural commodities in the tropical region and its fast achievements in low-carbon agricultural technologies, such as highly productive integrated crop–livestock systems, will reinforce its influence on world markets. So far, the country uses less than 50 percent of its geographic area as agricultural land. This fact, along with the possibility of using agricultural technologies to ensure environmental protection and expand production, will further strengthen the country's role in agricultural markets.

Unsurprisingly, Brazilian agriculture's success story has awakened the intense interest of other developing countries, mainly in Africa and in Latin America and the Caribbean, that want to know more about tropical agricultural technologies developed by Brazil. Responding to this increased demand, the Brazilian Agricultural Research Corporation (Embrapa) significantly expanded its participation in cooperation projects in other tropical regions in the last five years, broadening the possibilities for fruitful partnerships with other countries. The focus of Embrapa's cooperation is on technology transfer, capacity strengthening, and research cooperation.

Summit, launched in 2008, is paving the way for greater cooperation, such as through the transfer of agricultural technologies that meet the real needs of small-scale farmers in Africa. India is a leader in tropical technology—not only improved varieties but also resources management technologies, which are just as important for meeting farmers' needs. India is also an active player in the Interregional Initiatives for India, Brazil, and South Africa—which established the Facility Fund for Alleviation of Poverty and Hunger in Africa in 2003.

South Africa, itself a leader in agricultural technology, is a key player in the transfer of technologies to fellow African countries. In Latin America, Brazil has been actively involved in South–South cooperation for agricultural development, both within Latin America and in other developing regions, particularly Africa. In Africa, Brazil initially focused on the Portuguese-speaking countries of Angola, Cape Verde, Guinea-Bissau, and Mozambique, but the opening of Embrapa, Brazil's agricultural research agency, in Ghana in 2006 points to a new phase in its South–South cooperation. More recently, other African countries, including Benin, the Democratic Republic of the Congo, Ethiopia, Ghana, Guinea, and Kenya, signed technical cooperation agreements with Embrapa and began implementing joint projects.

THE PRIVATE SECTOR

The private sector is also taking on a larger role in agricultural research, poverty alleviation, and environmental sustainability. In January 2011, the World Economic Forum released an innovative road map for the agricultural development of its stakeholders.[8] The roadmap, developed by 17 global companies, was designed to leverage public- and private-sector investment; share environmental best practices; develop agricultural markets, including opportunities for small-scale farmers; and improve access to affordable and nutritious food. It represents an important milestone in the private sector's increased engagement in the global discourse on agricultural development and food security.

The private sector has now become one of the World Food Programme's top 10 donors. Furthermore, new emergency protocols that emphasize partnerships with the private sector were put in place to help the World Food Programme improve its ability to save lives and livelihoods in disasters and emergencies, most recently in Haiti, Pakistan, and the Horn of Africa. Another initiative, the Food Retail Industry Challenge Fund by the United Kingdom's Department for International Development (DFID), is supporting African farmers through innovative business partnerships. The fund aims to improve the lives of African farmers by increasing European imports of agricultural products from poorer countries in Sub-Saharan Africa. So far, the fund supports 11 innovative partnerships linking farmers with European markets.

In developing countries, the value chains of most food commodities are inefficient, with high transaction costs that lead to high food prices. At the G20 agriculture ministers' meeting on food security in Paris, participants discussed how the private sector could help stabilize global food markets and reduce price volatility. They jointly made a commitment to leveraging private-sector investment, using technology and information to stabilize global food markets and provide opportunities to poor farmers and consumers in the event of price spikes and volatility and extraordinary hunger levels. Their deliberations emphasized that the private sector can help to solve food insecurity, but that its activities must be conducted in collaboration with governments as part of an integrated strategy to make the global food system more sustainable.[9] The global leaders agreed on ways to better coordinate public- and private-sector efforts, including the formation of national-level partnerships to engage the private sector in sustainable agricultural development and the creation of a global forum to exchange best practices and provide inputs to the G20 on a regular basis.

PHILANTHROPIC ORGANIZATIONS

Private philanthropic and civil society organizations are promoting the global agricultural development agenda on a much greater scale than just a decade ago. Many international nongovernmental organizations are transforming themselves with new goals and approaches, by mobilizing resources for development programs, and by acting more independently from government-financed programs.

The private sector has now become one of the World Food Programme's top 10 donors.

The Bill & Melinda Gates Foundation has invested in an agricultural development program intended to help small farmers in Sub-Saharan Africa and South Asia boost their productivity, increase their incomes, and build better lives for their families. In the past decade (Figure 3), it has become an important donor to the Consultative Group on International Agricultural Research (CGIAR). As of June 2011, the Foundation had committed about US$12 billion to agricultural research and development, agricultural policies, and access to market systems, as well as financial services for the poor, water sanitation and hygiene, and policy advocacy.

Other philanthropic organizations have also emerged as major supporters of agricultural development, poverty alleviation, natural resource management, and risk management. For example, the Sir Ratan Tata Trust and the Navajbai Ratan Tata Trust in India are funding activities related to drought proofing, microfinance, and a revival of the Green Revolution. During 2010–11 the Trust allocated US$31 million, of which 75 percent was for rural livelihoods and communities.[10] The Howard G. Buffet Foundation has funded projects in more than 74 countries, including 32 African countries, on agriculture for nutrition. These projects are designed to benefit more than 1.5 million people by addressing poor crop yields, limited success with livestock, low incomes, and chronic hunger among vulnerable communities. The Foundation is also supporting global initiatives on conservation agriculture. Similarly, the PepsiCo Foundation (PepsiCo's philanthropic arm) is developing partnerships and programs to improve health, environment, and education in underserved regions.

FIGURE 3 **Top 10 donors to the CGIAR, 2000–10**

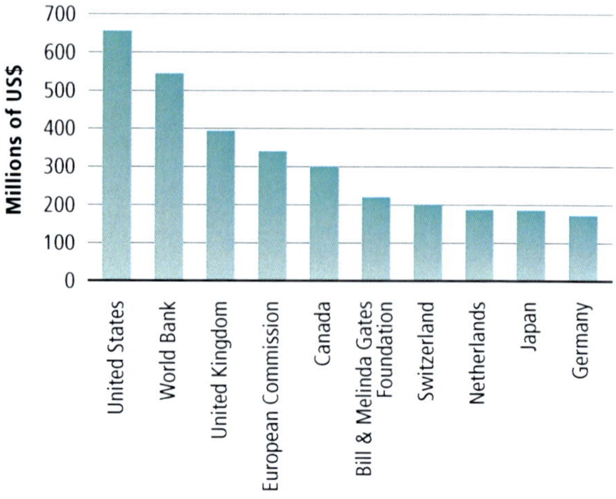

Source: Consultative Group on International Agricultural Research, *The CGIAR at 40 and Beyond: Impacts that Matter for the Poor and the Planet* (Washington, DC, 2011).

BOX 14

Private Philanthropy and Public Policy

Prabhu Pingali, Bill & Melinda Gates Foundation

The rise of private nonprofit organizations during the past fifteen years has transformed the nature of aid supply in a significant way. By 2011, the philanthropic sector had added to the number of organizations operating internationally and to total aid flows. Although no consolidated statistics exist, it is estimated that global private aid doubled between 2004 and 2009. Likewise, at US$52.5 billion in 2009, the value of private donations to developing countries may well have become comparable in scale to sector-allocated official development assistance.[1]

In the agriculture sector, the growth of philanthropic giving in parallel to the emergence of new bilateral donors—such as Brazil, Russia, India, China, and South Africa (BRICS), and Korea—comes at a critical time, when levels of investment in agricultural development remain largely inadequate. The significant withdrawal of donor support and national government attention to agriculture in the mid-1980s following the success of the Green Revolution left the global food system in a stagnant state. The resulting stagnation and decline in agricultural productivity growth has been felt throughout most of Africa and South Asia. The international community recently renewed its interest in agriculture following the 2007–08 and 2011 food price crises, and the trends in private and bilateral giving may signal the beginning of a new surge in international agricultural development and, ultimately, in improved food security worldwide.

Developments of 2011 continued into 2012, with Unilever launching a charitable foundation at the World Economic Forum 2012 with the goal of helping more than 1 billion people improve their health and well-being. It is also working with the World Food Programme's Project Laser Beam to help eradicate hunger and poverty in Bangladesh and Indonesia. It, in partnership with other organizations, has committed US$50 million over five years to create a replicable and sustainable solution targeted at the ultra-poor, especially women.[11]

The Rockefeller Foundation has reoriented its philanthropic mission to promote human well-being with greater focus on Africa. It launched the Alliance for a Green Revolution in Africa (AGRA) in partnership with the Bill & Melinda Gates Foundation in 2006. This Africa-based and Africa-led organization is charged with sustainably increasing the productivity and profitability of smallholder farms throughout Africa. It seeks to provide access to more resilient seeds that produce higher and more stable yields, promote soil health and productivity, build more efficient local, national, and regional agricultural markets, promote better policies, and build partnerships to develop technologies and institutional changes needed to achieve a green revolution. AGRA received a grant of US$5 million for 2011 and 2012 from the Bill & Melinda Gates Foundation and the Rockefeller Foundation to improve the productivity and incomes of small-scale farmers in Africa by integrating its programs with those of partners such as African governments, CGIAR centers, the private sector, and various network programs in the breadbasket regions of key countries.[12]

MOVING TOWARD A NEW DEVELOPMENT DYNAMIC

The rise of new players has fueled calls for new state and nonstate players to become even more involved in the governance of global food security. The G20, in particular, has filled a gap in global governance by creating coalitions that connect advanced and developing countries. For example, the G20 affirmed its support for a widening role for the Committee on World Food Security at its June 2011 meeting of agriculture ministers.[13] In its Ministerial Declaration, the G20 indicated its support for the ongoing work of the Committee as the

Beyond aid flows, the philanthropic sector has also changed the way in which aid to agriculture is being channeled within countries. International nongovernmental organizations and voluntary organizations have been able to deliver essential services and public goods, thereby assuming critical roles that governments or international donors cannot. Meanwhile, private foundations have focused on strengthening the capacity of local development institutions that can adapt solutions to local conditions. In addition, through investments at all levels of agricultural value chains, private foundations have catalyzed the development and piloting of innovative solutions, approaches, and models—from planting high-quality seeds and improving farm-management practices to streamlining methods of bringing crops to market.

The Bill & Melinda Gates Foundation has emerged in recent years, alongside major multilateral and bilateral donors, as one of the leading contributors to agricultural development aid. It supplies approximately US$400 million per year in agriculture-sector grants, with a particular focus on smallholder productivity growth in Sub-Saharan Africa and South Asia. Through 2011, the Foundation committed US$2 billion to its Agricultural Development Program, which has helped initiate action among partners at both (1) the global level, in support, for instance, of high-end agricultural research and development by the CGIAR, and (2) national and local levels, in direct support of farmers and the situations (including the knowledge, socioeconomic, and ecological systems) they operate within. Investments have ranged from the development of global public goods (such as improved crop and livestock varieties, farming practices, and agricultural data and statistics) to implementing and targeting successful programs (for example, through efforts to address local market failures or to ensure that improved tools reach the hands of farmers). The Foundation believes that these collaborative efforts will help enhance smallholder productivity and reduce poverty in Sub-Saharan Africa and South Asia.

BOX 15

The Private Sector and Food Security

Derek Yach, PepsiCo

In 2011, PepsiCo Inc. announced a partnership with the US Agency for International Development and the United Nations World Food Programme intended to create economic stability for smallholder chickpea farmers in Ethiopia. The partners are working together with Ethiopian farmers, local food manufacturers, research institutes, and donors to increase chickpea productivity among smallholder farmers, develop food products needed to reduce hunger, and, in time, build an export to improve livelihoods and also support part of PepsiCo's supply chain needs.

This is one example of how private companies can contribute to food security. PepsiCo, which has a large and growing chickpea-based hummus business, and other companies are constantly seeking ways to create new markets, invest in emerging economies, advance healthy nutrition, ensure environmental sustainability, and drive the long-term growth and profitability of their companies. These goals often overlap with the objectives of public organizations and others trying to end hunger and reduce poverty. The World Economic Forum, with its New Vision for Agriculture, has recognized the benefits of these public–private partnerships and is stimulating and developing multistakeholder programs in several countries.[1] The multistakeholder approach to global food policy appeals to food companies because it reduces risks they

foremost inclusive international and intergovernmental platform. In particular, it stressed the Committee's responsibility for enhancing engagement with the private sector and strengthening North–South, South–South, and triangular cooperation.

> The G20, in particular, has filled a gap in global governance by creating coalitions that cut across advanced and developing countries.

At its own summit in October 2011 in Rome, the Committee on World Food Security included both the private sector and philanthropic organizations at the table for the first time.

South–South cooperation is becoming part of the global agenda for aid effectiveness.[14] Triangular cooperation between traditional aid donors, emerging aid donors, and recipient countries is one way forward. Much of China's commitment to African agriculture is embodied in its donation of US$30 million in 2009 to the Special Programme for Food Security of the Food and Agriculture Organization of the United Nations. Many bilateral aid agencies, such as those of Germany, Japan, and the United Kingdom, are closing their bilateral programs in China but are actively exploring new partnerships with China to aid Africa. For example, after closing its bilateral program in March 2011, DFID China started to develop activities under the Global Development Partnership Programme, which is DFID's new framework to engage emerging powers and new partners in global development. The Programme will support collaborative activities with China in sectors such as agriculture, climate change, and health, some of which will target selected developing countries.

The Bill & Melinda Gates Foundation is also initiating its own cooperation with emerging economies in Africa.[15] In November 2011, the Foundation announced a partnership with the government of Brazil aimed at improving the agricultural productivity of small farmers in Sub-Saharan Africa and South Asia. A grant of about US$2.5 million was awarded to Embrapa to enable Brazilian and African agriculture research organizations to collaborate on agricultural development to improve the productivity of smallholder farmers in the developing world. At almost the same time, the Gates Foundation and the Chinese Ministry

share with other sectors, such as those related to climate change and the volatility of essential commodity prices, and with other private companies, such as the risk of entering new markets. Multistakeholder initiatives allow for deploying blended public and private funding sources in ways that meet the private sector's need for profitability and development agencies' need to enhance rural development and alleviate hunger. They also allow for scaled investments in infrastructure beneficial to business and society.

PepsiCo's involvement in such multistakeholder discussions and actions is transforming how the company does business and with whom it partners. For example, in a joint initiative with the Chinese Ministry of Agriculture, PepsiCo announced it would build and operate demonstration farms that use the most advanced irrigation, fertilizer, and crop management techniques.[2] PepsiCo demonstrated increase in potato yields in China to 45 tons per hectare, meeting the global standard, while achieving up to 50 percent reduction in water consumption in potato cultivation by implementing advanced irrigation techniques.[3] In Mexico, through a partnership with the Inter-American Development Bank and the government, the company has co-invested in building sunflower production capability that will reduce PepsiCo's reliance on palm oil and, through advance-purchase agreements, lift local farmers out of poverty. In India, PepsiCo works with local academics and consumer insights groups to increase young women's access to reasonably priced, iron-fortified, nutritious products that allow the company to reach poor urban communities and share messages that resonate with them.[4]

Companies such as PepsiCo have fresh perspectives and viewpoints useful in the fight against hunger and poverty. However, the private sector does not have all the answers. Partnerships, collaboration, and knowledge exchange between the private and public sectors, as well as civil society, are what will truly help solve development challenges and benefit communities worldwide.

of Science and Technology signed an agreement to produce innovative technologies to boost the progress of developing countries and promote the achievement of the Millennium Development Goals. These partnerships demonstrate the critical role that emerging economies like Brazil and China can play in driving innovation to reduce world poverty and hunger.

Still, the opportunities presented by these new players have not been fully harnessed.[16] There seems to be a consensus that increased investment in agriculture should give priority to smallholder food production, yet the New Vision for African Agriculture launched by 17 multinational corporations at the 2011 World Economic Forum made no mention of smallholder farmers' organizations or the Committee on World Food Security.[17] At the same time, the private sector has, until recently, been largely absent from the reformed Committee on World Food Security, and its presence has been essentially limited to multinational corporations, although new forms of private sector participation have been proposed.[18] To involve new players and retain traditional players in the global food security system, it is essential to strengthen collaboration and build trust among different stakeholders through the establishment of strong coalitions of willing partners at the local, regional, and global levels. ■

REGIONAL DEVELOPMENTS

Food Policy Taking Shape at the Local Level

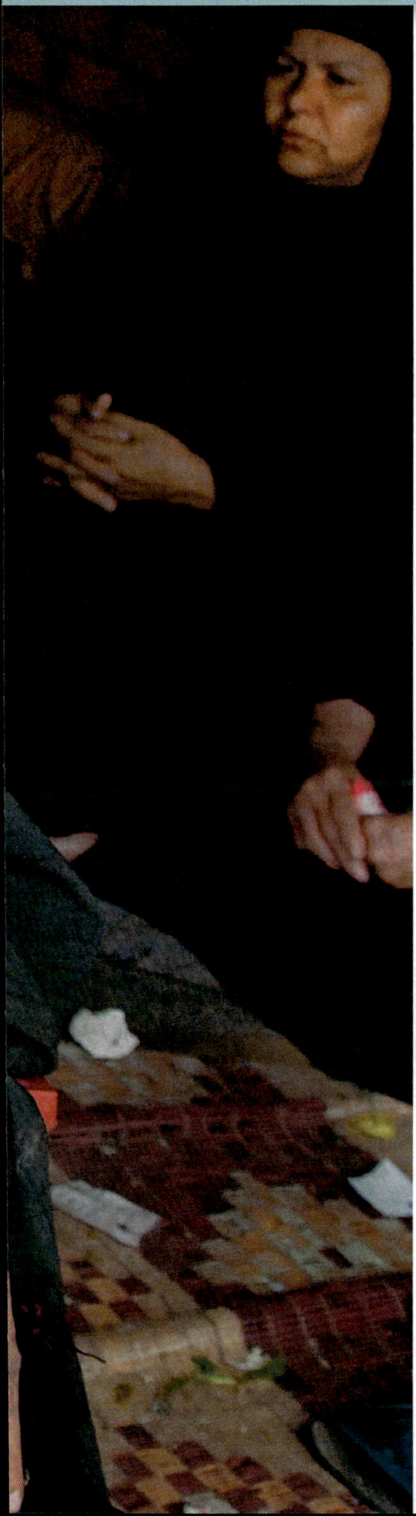

APART FROM 2011'S GLOBAL CHALLENGES AND OPPORtunities for food policy, important developments with potentially wide repercussions took place in individual regions and countries, which are discussed in this chapter. As 2011 opened, events in Tunisia ignited the "Arab Awakening" in North Africa and the Arab world, posing new challenges for food security in the region. To the south, Sub-Saharan Africa made progress on its continent-wide framework for raising agricultural growth to improve food security and reduce poverty. That region also pushed forward with efforts to increase agricultural productivity and incomes while pursuing economic transformation beyond agriculture. Therefore, although a food crisis struck the Horn of Africa, progress elsewhere belies the image of a continent mired in gloom. India, having already achieved economic growth and higher agricultural productivity, turned its attention to domestic food security. It responded to an extraordinary right-to-food movement by implementing the world's largest antihunger program and revived other initiatives designed to overcome malnutrition, which remains high despite the country's economic success.

Food security at home was also the primary goal of China's major 2011 food policies and investments, which focused on water conservation and agricultural research and development. China also engaged in agricultural development abroad, particularly in Africa—though its engagement is smaller and of longer standing than often believed (see Box 12 in Chapter 8). Brazil has been even more proactive in increasing its role in the global food system. By successfully expanding agricultural production, the country has solidified its position in world agricultural markets and become an increasingly active player on the international scene (see Box 13 in Chapter 8).

Middle East and North Africa

The Arab Awakening and Food Security

Clemens Breisinger, Olivier Ecker, Perrihan Al-Riffai, and Bingxin Yu, IFPRI

The self-immolation of a young Tunisian man in December 2010 sparked a wave of protests and uprisings across the Arab world in 2011. The Arab Awakening has most directly affected Bahrain, Egypt, Libya, Syria, Tunisia, and Yemen, but it may also provide impetus for change in other Arab countries. But with opportunities have come significant challenges for governments and civil society. In the short run, countries in transition have seen a sharp slowdown in economic growth and a related rise in unemployment. In addition, political fluidity, coupled with rising food and fuel prices, has led to widespread increases in food and fuel subsidies, public-sector wages, and other government welfare spending.[1] As a result, oil-importing countries in particular face growing budget gaps, inflation in both oil-exporting and oil-importing countries is expected to rise, and foreign exchange reserves are expected to contract. Moreover, conflict and security concerns have dampened tourism, remittances, and foreign direct investment throughout the region.[2] Mastering these short-term challenges will be critical for a successful transition process.

Achieving prosperity will also require addressing long-standing challenges in the region. Not only political issues, but also socioeconomic issues likely contributed to the onset of the Arab Awakening.[3] These socioeconomic factors include unemployment, especially among the youth, growing income disparities, and a high risk of food insecurity for both countries and households (Figure 1). While public spending is relatively high in Arab countries, it contributes less to economic growth than it does in other world regions. And economic growth does not trickle down to the poor in Arab countries as readily as it does in the rest of the world. In most countries, agricultural growth, for instance, leads to substantial reductions in child undernutrition, but not in the Arab countries. Manufacturing and service sector–led growth in the Arab region is more in line with the rest of the world, and, given a conducive business environment, these sectors have large potential to create jobs and drive improvement in the well-being of the poor.

Finding solutions to these challenges will require a broad economic road map and country-specific development strategies. Research can help with evidence that identifies the pros and cons of alternative policy options. To improve food security and reduce poverty, Arab policymakers should focus on the following key areas:

1. **Improve data and capacity for evidence-based decisionmaking.** Decisions based on flawed data can damage economies, harm people's well-being, and lead to significant financial losses, so policymakers and voters need realistic and accurate baseline data. Currently, some official numbers, such as official poverty and inequality indicators, may underestimate the extent of the problems facing many Arab countries. Admitting that these data are flawed will help increase the credibility of policymakers and allow them to set targets against which future progress can be more realistically assessed. Countries should not only improve the quality of their data, but also make that data more accessible.

2. **Foster growth that enhances food security.** Fostering economic growth is fundamental for enhancing food security, yet certain types of growth do more to improve food security at the country level, whereas other types of growth have stronger effects on household-level food security. Export-led growth generates foreign exchange revenues for food imports and thus improves food security at the country level. Inclusive growth that generates jobs and increases incomes for the poor enhances food security at the household level. Growth, combined with appropriate tax systems, also

FIGURE 1 The risk of food insecurity in the Arab countries plus Iran and Turkey, 2011

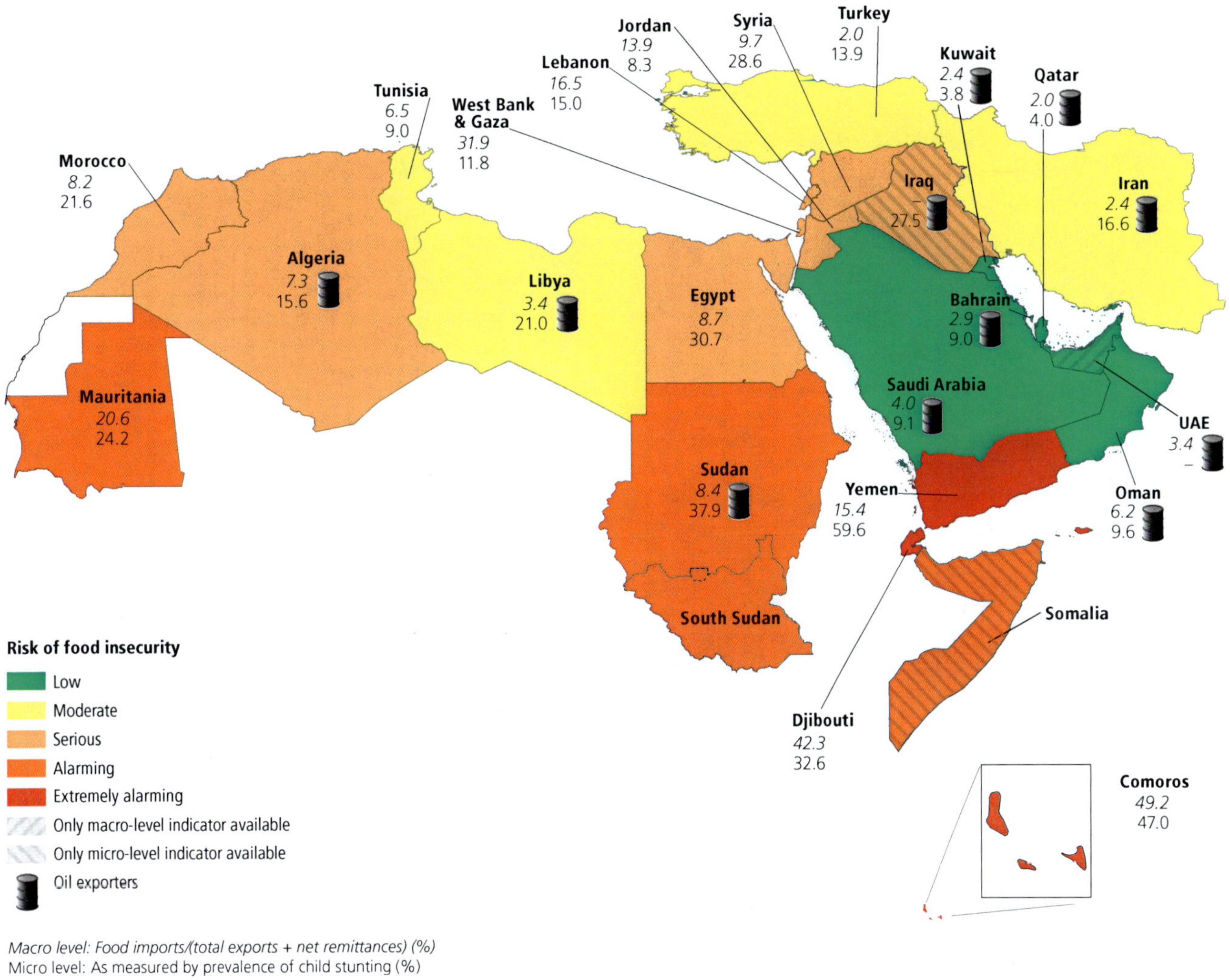

Source: C. Breisinger, O. Ecker, P. Al-Riffai, and B. Yu, *Beyond the Arab Awakening: Policies and Investments for Poverty Reduction and Food Security,* IFPRI Food Policy Report (Washington, DC: International Food Policy Research Institute, 2012). **Note:** The risk of food insecurity describes the vulnerability of countries (macro-level) and households (micro-level).

generates government revenues that can be directed to public spending on food security.

3. **Improve the efficiency and allocation of public spending.** Arab countries urgently need to revisit the efficiency and allocation of public spending and make improvements where necessary. In general, most oil-exporting countries with fiscal surpluses can compensate for inefficiencies by increasing spending, whereas oil-importing countries with fiscal deficits must reallocate spending and improve spending efficiency to achieve development goals. Both oil importers and oil exporters, however, should have a genuine interest in achieving higher returns on their spending. A wide array of country-specific factors affects public expenditure efficiency, such as the level of economic development, the size of the public sector, public-sector competence, governance, political stability, and security of property rights. As in the case of growth strategies, investment plans

have to account for country-specific conditions. These investment plans should also be aligned with growth strategies at the subsector and subnational levels to ensure maximum consistency and development impact.

Designing and implementing these and other policy and investment priorities will require visionary leadership, sound laws and institutions, politicians who are accountable and listen to the voices of the people, and a civil society that is patient and accepts the tenets of democracy. The Arab world has awakened. It is now time to take the steps necessary for a food-secure Arab world without poverty.

Africa

Moving beyond Recovery to Economic Transformation

Ousmane Badiane, IFPRI

After decades of stagnation, African economies embarked on a remarkable recovery that started in the late 1990s and continues today. As a region, Sub-Saharan Africa is growing at average rates exceeding 5 percent annually.[4] The next challenge for African countries is to move from recovery to economic transformation (see Box 16). Such a transformation will require countries to raise productivity in the agricultural and rural sectors while diversifying into higher-productivity, urban-based manufacturing and service sectors that produce higher-value goods and services. The result would be greater overall economic productivity and progress in lifting rural incomes toward the level of urban incomes.

UNBALANCED GROWTH

During most of the period preceding the current recovery, overall productivity in African countries has stagnated at best. This happened because labor migrated from the relatively better-performing agricultural sector into the nonagricultural sector, which is dominated by the less productive yet rapidly expanding service sector.[5] Major factors behind this process have been the relative neglect of agriculture in national polices and the rapid

BOX 16

Pushing for Progress on Agriculture in Africa

Tsitsi Makombe, IFPRI

The year 2011 brought significant progress in the implementation of the Comprehensive Africa Agriculture Development Programme, the African Union's continent-wide framework guiding national efforts to increase agricultural growth and progress toward poverty reduction and food and nutrition security. Six countries held Programme roundtables and signed compacts, bringing to 29 the number of countries that have done so. Most countries have now adopted a national agricultural investment plan following a technical review of the plan and a business meeting that validated the plan and determined how it would be financed.

In the national agricultural investment plans, countries have pledged to raise their agricultural investment level. The Programme asks governments to increase agricultural spending to 10 percent of national budgets and achieve an annual agricultural growth rate of 6 percent. Based on the latest available data, only 8 countries have reached the 10 percent budget target, while 17 attained growth rates of at least 6 percent in 2009.[1] More countries are expected to move toward the budget target following their adoption of national agricultural investment plans. For example, Rwanda's agricultural budget share was just 3.5 percent in 2007 when the country signed its compact. But after Rwanda completed a national agricultural investment plan in 2009, the country's agricultural budget share rose to 6.8 percent by 2010–11.[2]

pace of urbanization. Because the share of people employed in agriculture has fallen much faster than their productivity has risen, agriculture as a percentage of gross domestic product (GDP) has declined rapidly in most African countries. In most of these countries, labor has migrated out of agriculture faster than it can be absorbed into the nonagriculture sector, undermining productivity growth—not only in agriculture, but in the economy as a whole. The problem is as much slow growth in agriculture as it is poor labor absorption outside agriculture.

For the average African country, the size of the agriculture sector as a share of GDP is nearly 20 percentage points smaller than one would expect based on the actual level of economic development. The service sector, on the other hand, is about 20 percentage points larger than one would expect.[6] This imbalance in the growth of the two sectors has slowed productivity and income growth across Africa. How can African countries capitalize on the ongoing agricultural and economic growth recovery of the past 15 years to boost the transformation of their economies?

NOT JUST AGRICULTURE, BUT INDUSTRIALIZATION

To achieve successful structural change, countries need to produce more sophisticated, higher-value goods for which demand expands globally as incomes rise around the world. The decades of economic stagnation preceding Africa's current recovery meant that African economies failed to promote product sophistication. From 1962 to 2000, the average indicator of the degree of sophistication of all products produced by African countries was 50 to 60 percent lower than the estimated average for a sample of 97 other countries.[7] The value of the estimated indicator for agricultural products not only is low, but has stagnated since the 1960s.

During the 1960s, most African countries pursued an industrialization strategy led by the public sector and designed to substitute for imported goods. The widespread failure of these efforts has left most African countries without any bold and credible strategies for industrialization.

Successful economic transformation is associated with a series of changes in the overall economy. These changes include a rise in the absolute

Donors intensified their Programme alignment and harmonization efforts in 2011. They funded the Programme bilaterally and multilaterally through its Multi-donor Trust Fund and Global Agriculture and Food Security Program. The Multi-donor Trust Fund supports Programme implementation processes while the Global Agriculture and Food Security Program addresses potential underfunding of national agricultural investment plans. With a US$46.5 million grant in 2011 to support Liberia's national agricultural investment plan, the Global Agriculture and Food Security Program has now allocated more than US$270 million to support the national agricultural investment plans of six African countries.

The Programme implementation process has had its challenges. Engagement of civil society and the private sector has been limited in some countries while some national agricultural investment plans have had unrealistic growth and poverty reduction targets and funding requirements. Nonetheless, these challenges are already being addressed. The New Partnership for Africa's Development (the African Union's planning and coordinating agency) and development partners commissioned guidelines for nonstate-actor participation, and the 7th Comprehensive Africa Agriculture Development Programme Partnership Platform meeting, held in March 2011, called for their systematic and targeted dissemination across countries. Efforts are also underway to strengthen country implementation capacity and improve the quality of national agricultural investment plans. For example, Strategic Analysis and Knowledge Support Systems in individual countries—which are being established with the technical support of the International Food Policy Research Institute and Regional Strategic Analysis and Knowledge Support Systems—are expected to improve analytical and review capacities for evidence-based agricultural policy planning, implementation, and monitoring and evaluation. It is noteworthy that Programme implementation coincides with the continent's agricultural growth recovery.[3] Renewed agricultural growth provides a solid foundation for African countries as they work to implement their agricultural investment plans.

and relative size of industrial output; an increase in the number and nature of industrial goods; new techniques for combining labor, capital, and technology to produce goods in industry, as well as changes in the related costs; and a change in the sources supplying the economy with existing and new products.[8] In this process, industrial enterprises need to learn how to combine and recombine existing and new assets to establish new businesses and create products to address new markets.[9]

The risks and uncertainties faced by entrepreneurs in producing a new good for the first time are a major determinant of an economy's capacity to diversify into higher-productivity goods. As they increase their investments in agricultural development,[10] African countries also need renewed industrialization strategies to build on the current recovery. Such strategies should target the technological, institutional, and infrastructural factors that raise the level of risk and uncertainty related to entrepreneurial innovation. Ultimately, industrialization policies should expand a country's arsenal of technologies and its ability to apply them to create new, higher-valued goods.[11] African countries will need to rediscover ways of stimulating industrial growth and may need to look at emerging Asian countries, where public action in support of industrial growth has been a central element of economic development.[12]

Africa's industrial policies should seek to encourage the creation and growth of enterprises, not just in industry, but also in agribusiness and the informal sector. In the first three decades of the 21st century, demand for food in Africa is expected to grow by US$100 billion, of which one-third could be met by smallholders.[13] This rising demand creates a real incentive and opportunity for industrialization based on agribusiness in the medium term. Strategies should include not only programs to raise productivity on farms, but also development of new processing and packaging technologies to support product innovation and creation of competitive distribution networks and transport infrastructure that cross national boundaries. The Comprehensive Africa Agriculture Development Program can be helpful here by promoting evidence-based policy planning and implementation and creating opportunities for strategic public-private partnerships and business-to-business alliances. Industrialization policies should also focus on the informal sector currently producing low-quality household goods. This sector has huge potential for enterprise growth and consolidation, as well as product improvement and innovation.[14]

India

Enshrining the Right to Food

M. S. Swaminathan,
M. S. Swaminathan Research Foundation

Hunger and malnutrition have long devastated the people and progress of India. In 2011, the Indian government took a major step toward putting an end to this injustice nationwide (see Box 17). With the introduction of the National Food Security Act, India's "ship-to-mouth" existence (as the country's previous reliance on food shipments is commonly termed) is en route to one of food security and health through the implementation of the world's largest social-protection program against hunger using homegrown food.

In light of the conditions that have prevailed in India since its independence in 1947, the National Food Security Act is nothing short of historic, as it positions institutions and individuals to radically reduce hunger and malnutrition. In the recent past, specifically during the 1960s, India received the highest amount of concessional food aid worldwide; in 1966 alone, the country imported more than 10 million tons of wheat. Today, based on the requirements of the new National Food Security Act, India is due to commit more than 60 million tons of homegrown wheat, rice, and millet at a highly subsidized price to citizens living below the poverty line. This is only one of the numerous ways that this new Act (once it is carefully reviewed, slightly amended, and enacted as law) promises a multi-faceted attack on hunger.

Despite India's countless efforts to increase food and nutrition security in the past, widespread improvements have been minimal. A

BOX 17

India's Economy Roars Ahead but Nutritional Improvement Is Stalled

P. K. Joshi and Suneetha Kadiyala, IFPRI; S. Mahendra Dev, Indira Gandhi Institute of Development Research

India allocated additional resources in 2011 to ongoing programs aimed at reducing poverty and ending hunger and malnutrition. The country has high economic growth, bumper foodgrain production, and substantial food buffer stocks, which would usually indicate improvement in poverty and malnutrition numbers. However, poverty in India remains pervasive, malnutrition is widespread, and the country's nutritional status lags far behind Brazil or China. To address this, the Government of India initiated, and is now supplementing, an array of programs and policy measures for the welfare of the poor, especially women and children.[1]

Currently, more than 40 percent of India's population earns less than US$1.25 a day. Estimates show that one-third of all Indian women are underweight; almost half of Indian children are stunted and 40 percent are underweight; and rates of micronutrient deficiencies are also extremely high. The key reasons for India's nutrition numbers lagging so far behind its economic numbers are (1) low agricultural productivity; (2) low incomes for a majority of the rural population; and (3) poor people's limited access to education and food, health, and nutrition programs.[2] To boost incomes, among other concerns, the government's budget outlay for the Mahatma Gandhi National Rural Employment Guarantee Scheme increased from US$2.60 billion in 2006–07 to US$8.91 billion in 2010–11. This flagship program aims to enhance livelihood security of households in rural areas by providing at least 100 days of guaranteed wage employment in a year and mandates 33 percent participation by women.

The Scheme has generated employment opportunities in rural areas and has had a positive impact on consumption expenditure, intake of energy and protein, and asset accumulation by the rural poor. The government intends to expand the scope of the Mahatma Gandhi National Rural Employment Guarantee Scheme by providing employment that creates permanent assets for the country (such as water conservation structures) and builds infrastructure (such as rural roads). To improve access to food, the public distribution system provides highly subsidized food to the population below the poverty line. The public distribution system is expected to be reoriented and implemented through the proposed Food Security Act. Implementation of the Act was expected in 2011 but is currently awaiting parliament approval. Many observers see mixed results from these programs, driven mainly by the variable quality of program governance and lack of appropriate coordination by different departments. The challenge is to make these programs efficient and more accessible to the poor, or effective at scale.

India has a long way to go to ensure nutritional security. Some estimates suggest that with the current pace of actions, the target of achieving the first Millennium Development Goal to halve the number of undernourished people by 2015 will be achieved by India only in 2042—a 28-year lag.[3] This is unacceptable given the country's outstanding economic performance. Programs and resources require convergence, transparent institutions, and inclusive governance for greater impact. Continued poverty and undernourishment would undermine India's progress as a high-performing emerging economy.

The pathways between agriculture and nutrition are now well recognized (see Chapter 6), which is why existing programs need to dovetail with agricultural improvements. A disconnect exists between agricultural production and nutritional security when, as in India, the majority of farmers holding less than two hectares are net buyers of food. Production of high-value and nutritive commodities—such as milk, poultry, meat, and fish—by smallholders would augment their income and make them food and nutrition secure.

recent report states that "maternal and child undernutrition levels remain persistently and unacceptably high."[15] The 2011 edition of the Human Development Report points out that body mass index, which is a good measure of an individual's nutritional status, has remained low for the past twenty-five years.[16] But, how can this be so, when innovative programs designed to deliver nutrition, education, and healthcare in integrated ways (including the Integrated Child

Development Service, the Mid-Day Meal Programme, and other programs initiated by international agencies like UNICEF and the World Food Programme) exist nationwide?

The answer is a lack of convergence and synergy between programs and across platforms. The Food Security Act must therefore ensure that programs address all essential components of food security. The draft of the bill introduced to Parliament in December 2011 focused only on economic access to food without making explicit reference to the other essential components of food security, including access to clean drinking water, sanitation, primary healthcare, and nutritional education. In its current form, the bill also neglects to mention the role of farmers or of public participation, although the *panchayat raj* institutions (whose members are elected by residents of their villages) and particularly *gram sabhas* (which provide a meeting where residents can raise concerns) can play a vital role in ensuring the effective and non-corrupt administration of legal entitlements.

To achieve widespread sustainable food and nutrition security, the draft legislation of the National Food Security Act will need to be carefully reviewed and strategically revised. The lifecycle approach to nutrition support should be embraced, and women should be granted legal entitlement as heads of household so that they are equally empowered to make financial and food-related decisions for their families. The public distribution of food should be strengthened, and the food basket within it should be expanded beyond wheat and rice to include nutritious cereals like jowar (sorghum), bajra (pearl millet), ragi (finger millet), and maize. The Public Distribution System often sees difficulty with targeting, which can be a major source of corruption. In states like Tamil Nadu and Kerala, governments may be unwilling to shift from their successful universal public distribution systems to a targeted national system, so an efficient system must have well-defined exclusion criteria and promote a culture of honesty (since those who do not need social protection from public funds would not demand highly subsidized food). In 2012, we hope to see the world's largest social protection scheme begin to enable India to reap a demographic dividend from its youthful population.

With the right modifications and the will to carry it out, the Food Security Act will also create a win-win situation for both resource-poor farmers and consumers. Procurement at a remunerative price will be the greatest stimulus for increasing farm productivity and production. For this reason, substituting cash for grains, while tempting, must be avoided; currency can be printed, but grains can be produced only by farmers—who constitute nearly 700 million of India's 1.2 billion people. It is the duty of a democratic society to ensure the well-being and survival of all people, and food security for all will help to assure income security for farm families.

China

Feeding a Billion People

Jikun Huang, Center for Chinese Agricultural Policy, Chinese Academy of Sciences

China's most significant food policy development in 2011 was its decision to put food security firmly at the top of its list of concerns by making three important policy decisions. First, the nation's most important policy document—the Number 1 Document—laid out plans to invest about US$630 billion in water conservancy in the next 10 years to combat increasing water scarcity. These plans include implementing institution and policy reforms to improve water-use efficiency.

Second, China has released the 2012 Number 1 Document specifically focused on innovation in agricultural science and technology and boosting agricultural productivity. The annual growth rate of public spending on agricultural research and development in real terms increased from an average of 16 percent from 2000–09 to more than 20 percent in 2010–11 and is expected to grow in the coming years.

Third, in November 2011 the national poverty standard, measured as annual per capita income in rural areas, was set at 2,300 yuan (about US$365), almost double the poverty standard in 2009. This is estimated to bring 13.4 percent of the rural population (or 128 million people) under the poverty alleviation program, which will have significant implications for poverty reduction and household food security.

China's ability to feed a population growing in both size and income has been impressive considering the country's natural-resource constraints. China supported more than 20 percent of the world's population and achieved nearly 98 percent overall food self-sufficiency in 2010 despite per capita water availability that is only one-fourth of the world average and arable land that accounted for only 8 percent of the world total. China's grain production grew for seven consecutive years from 2004 to 2010, and the government estimates that grain production surpassed 570 million tons in 2011, exceeding the record set just a year earlier.

China's agricultural success in the past three decades has contributed significantly to improved national food security. The driving forces of this success include China's household responsibility system, which distributed land equally to rural households; the application of science and technology to agriculture; investing in agricultural land and water; and market reform. However, it is hard to see how these successes can continue without significant innovation. For example, as Chinese citizens have seen their incomes increase, their demand for meat has also increased. This led China to shift from being a net exporter of maize—which is used as feed for farm animals—to a net importer in 2010. Given China's natural resource constraints, the country will likely become a more important maize importer in the near future, making its food security more dependent on other countries. It is clear that while China's accomplishments are impressive, great food-security challenges remain. ■

FOOD POLICY TOOLS AND INDICATORS

TO DEVELOP AND IMPLEMENT EFFECTIVE FOOD POLICIES, DECISIONMAKERS need resources, institutional capacity, political will, solid evidence, and timely information, among other things. As part of IFPRI's mission to find sustainable ways to reduce poverty and end hunger and malnutrition, the Institute develops and shares global public goods, including tools and indicators intended to help decisionmakers address real-life questions—ranging from models and networks to datasets and indexes. The tools include interactive ways to measure or conceptualize data; the indicators are standardized measurements that signify thresholds for certain aspects of development. While the content that can be extracted from these knowledge products is useful in and of itself, its practical purposes are enhanced and better understood by the various online networks that IFPRI co-facilitates that connect users to each other and provide a platform for discussion. These resources also provide benchmarks for monitoring and evaluating policies and programs to assess what works, what doesn't, and where improvements can be made.

Food Policy Tools

AGRODEP: AFRICAN GROWTH AND DEVELOPMENT POLICY MODELING CONSORTIUM

As African countries strive to increase agricultural productivity and attain greater economic growth, the policy reform process is fueling the demand for policy-relevant research. Meeting this demand requires strengthened local capacities in economic modeling, targeted partnerships, and broadened linkages to the ongoing policy debate. The African Growth and Development Policy (AGRODEP) Modeling Consortium was developed to help meet this demand. AGRODEP's goals are to mobilize a critical mass of modeling experts within Africa's policy research and analysis community, broaden access to innovative research methodologies, and facilitate partnerships with the global research community.

AGRODEP aims to position African experts—rather than external actors—to take a leading role in the study of strategic development questions and the broader agricultural growth and policy debates facing African countries. Started in 2010 by IFPRI and several partners, AGRODEP is building a network of African researchers with both the skills and the modeling and data infrastructure to lead this effort. AGRODEP shares core economic models; establishes a central online database for statistical, economic, and geospatial data; and combines a network of experts with a community of practitioners. All of this allows researchers to tackle major policy questions facing African countries strategically.

The AGRODEP website, launched in October 2011, serves as a portal for economic data and models and offers related resources, such as data-visualization tools. Event announcement pages facilitate networking activities among AGRODEP members, staff, partners, and governance groups.

Website: www.agrodep.org
Contacts: Ousmane Badiane and Betina Dimaranan
Email: info-agrodep@agrodep.org

FOOD SECURITY PORTAL

The price of food has far-reaching effects, having an impact on everything from poverty and nutrition to trade and markets to social and political unrest. This makes access to accurate, relevant, and useful global price and market information more important than ever. The Food Security Portal is a unique tool that provides users—including researchers, decisionmakers, and the media—with a one-stop resource for the latest food security information.

The Portal informs the global food security discussion with breaking news; regularly updated data from leading organizations like the UN Food and Agriculture Organization, the World Trade Organization, and the US Department of Agriculture; and interactive data-analysis tools. Policymakers can use the information and tools to track and analyze food security and price trends, enabling them to enact more informed, effective policies.

The Food Security News Hot Spot heatmap provides instant visualization of countries with the most food-security news items on a given day. The data API in the developer toolkit contains data for more than 40 indicators related to food security, commodity prices, economics, and human well-being. Users can download this data to track calorie supplies per capita, percent of undernourished children, consumer price indexes, global inflation, poverty rates, exports, imports, population information, and more.

One of the newest tools in the Food Security Portal is the Excessive Food Price Variability Early Warning System. This one-of-a-kind tool provides a visual representation of historical periods of excessive global price volatility from 2000 to the present, as well as a daily volatility status. This status can alert policymakers when world markets are experiencing a period of excessive food price volatility. The information can then be used to determine appropriate country-level food-security responses, such as the release of physical food stocks. This tool supports two recommendations of the 2011 G20 meetings. First, it provides information to the Agriculture Market Information System (AMIS) on price variability in global markets and identifies the presence and duration of excessive price variability. Second, it supports the design of emergency humanitarian food reserves coordinated by the World Food Programme (WFP).

Website: www.foodsecurityportal.org
Contact: Sara Gustafson
Email: s.gustafson@cgiar.org

FOOD SECURITY CASE MAPS

The Food Security Climate, Agriculture, and Socio-Economic Maps software (CASE Maps) is an interactive mapping tool policymakers and researchers can use to view potential changes in agricultural supply, demand, and food security based on different economic and biophysical drivers over time and across the globe. These interactive maps are based on data generated for *Food Security, Farming, and Climate Change to 2050*, published by IFPRI in 2010. CASE Maps display the exogenous model drivers and plausible future outcomes based on IFPRI's International Model for Policy Analysis of Agricultural Commodities and Trade (IMPACT) model.

The CASE Maps translate numerical results into a range of graphic representations, including scatter plots, bar charts, time-series graphs, histograms, and other basic statistical outputs. Those interested in climate change and its potential effects on food security in developing countries can view several indicators, including kilocalories available per capita per day, share of the population at risk of hunger, and the number of malnourished children age five or younger. CASE maps currently provide indicators for 15 different scenarios using 3 overall population and income futures, each with 5 different climate scenarios. Users can download the indicators by country and region. The scenarios provide a plausible range of outcomes based on the current policy environment. This information can then be used as a starting point for thinking about possible policy and program changes.

Website: www.ifpri.org/climatechange/casemaps.html
Contacts: Gerald Nelson and Amanda Palazzo
Email: ifpri-climatechange@cgiar.org

HARVESTCHOICE

The demand for tools to help prioritize and target cost-effective interventions with the greatest impact on food security is increasing among policy and investment analysts and the decisionmakers they support. HarvestChoice is responding to this demand by generating knowledge products that help guide strategic investments meant to improve smallholder livelihoods in Sub-Saharan Africa through more productive, profitable, and sustainable farming. HarvestChoice's hallmark approach is a spatially explicit data and evaluation framework designed to better represent the significant heterogeneity of farming systems and farm households across the region.

HarvestChoice's online resources allow users to explore data, apply tools, and create intervention-oriented maps that best address their own policy and investment questions. HarvestChoice tools can extract information from multiple fine-resolution data layers selected by the user. By selecting demographic, market-access, and crop-production layers, for example, users can tabulate population, travel time to markets, and crop-area attributes of any geographic area they specify.

The Agricultural Domain Visualizer/Reporter (ADVizR) allows users to map, tabulate, and chart a broad range of subnational agricultural development indicators according to selected geographic domains (for example, districts, agroecological zones, watersheds, farming systems, or marketsheds). Users select, in sequence, their target Sub-Saharan African subregion or country, as many as 20 development indicators of interest, and the specific geographic domains by which the indicators should be reported.

One of the HarvestChoice interfaces provides access to a suite of spatial data exploration and reporting tools (including the core functions of ADVizR). The website also offers options for third parties to embed these tools in their own sites.

Website: www.harvestchoice.org
Contact: Maria Theresa Tenorio
Email: m.t.tenorio@cgiar.org

RESAKSS: REGIONAL STRATEGIC ANALYSIS AND KNOWLEDGE SUPPORT SYSTEM

The Regional Strategic Analysis and Knowledge Support System (ReSAKSS) was established in response to the growing demand for credible information and analysis during the design of the Comprehensive Africa Agriculture Development Programme (CAADP). As a knowledge management platform, it offers smallholders, researchers, policymakers, and development professionals easily accessible data, analysis, tools, and research evidence. ReSAKSS is intended to promote dialogue and facilitate the review, learning, and adoption of best practices associated with the CAADP agenda and increase awareness of agriculture's role in poverty reduction and food and nutrition security.

ReSAKSS operates as a multicountry network of collaborating partners represented by regional web-based platforms. It is organized around three main activities: (1) *strategic analysis*: providing data, tools, and analysis for monitoring key growth and poverty-reduction priorities and assessing the impacts of interventions to address key questions, such as whether and how the interventions are having their desired impact on raising growth and reducing poverty and malnutrition; (2) *knowledge management*: developing and maintaining a common pool of up-to-date information on key indicators at various levels to support policy planning, analysis, and dialogue; documenting lessons from research, policy analysis, impact assessment, and practical experiences for improving future growth and poverty-reduction strategies; and facilitating access to knowledge products to support assessment, review, and benchmarking of growth and poverty-reduction strategies; and (3) *capacity strengthening and policy communications*: collaborating with national and regional networks to carry out strategic analysis while exchanging the valuable skills, training, and practical experiences that can strengthen local capacities and enhance the communication of key issues, findings, and strategies with policymakers and other stakeholders.

Website: www.resakss.org
Contact: Samuel Benin
Email: s.benin@cgiar.org

Food Policy Indicators

ASTI: AGRICULTURAL SCIENCE & TECHNOLOGY INDICATORS

Policymakers are increasingly recognizing greater investment in agricultural research and development as an essential element in increasing agricultural productivity. Data on the size and scope of research and development capacity and investments, together with the changing institutional structure and functioning of agricultural research agencies, enhance our understanding of how agricultural research and development promotes agricultural growth. Indicators derived from such information allow the performance, inputs, and outcomes of agricultural research and development systems to be measured, monitored, and benchmarked.

The Agricultural Science & Technology Indicators (ASTI) initiative is one of the few sources of information on agricultural science and technology statistics for low- and middle-income countries. ASTI provides comprehensive quantitative and qualitative information and subsequent analyses on investment, capacity, and institutional trends in agricultural research and development. (See Table 1.) ASTI data and analyses assist research and development managers and policymakers in improved policy formulation and decisionmaking at national, regional, and international levels. All outputs are made available on the ASTI website.

Website: www.asti.cgiar.org
Contact: Nienke Beintema
Email: asti@cgiar.org

TABLE 1 Public sector agricultural research and development (R&D) spending and staffing, by country

Country	Latest data available	Public research spending: 2005 PPP dollars (millions)	Public research spending: 2005 US dollars (millions)	Public spending as a share of ag. GDP (%)	FTE public researchers	Researchers per million economically engaged population in agriculture	Women as share of total researchers (%)
SUB-SAHARAN AFRICA							
Benin	2008	21.6	9.0	0.57	115.4	13.26	16.9
Botswana	2008	19.0	9.0	4.32	97.4	63.00	29.8
Burkina Faso	2008	19.4	7.4	0.43	239.9	3.05	11.6
Burundi	2008	9.6	3.0	1.78	97.8	2.64	14.8
Congo, Republic of	2008	4.6	2.3	0.85	93.8	9.35	16.2
Côte d'Ivoire	2008	42.6	23.2	0.54	122.6	13.92	16.8
Eritrea	2008	3.0	1.2	0.45	121.9	2.07	32.0
Ethiopia	2008	68.6	17.8	0.27	1,318.3	2.24	6.8
Gabon	2008	1.6	0.8	0.20	61.4	8.37	22.4
Gambia, The	2008	2.5	0.7	0.50	37.7	4.38	13.7
Ghana	2008	95.4	39.2	0.90	537.1	16.48	17.0
Guinea	2008	4.0	1.3	0.18	229.2	1.05	3.2
Kenya	2008	171.5	67.0	1.30	1,011.5	13.36	26.8
Madagascar	2008	11.9	3.9	0.27	212.4	1.79	29.8
Malawi	2008	21.4	7.1	0.68	126.5	4.35	15.7
Mali	2008	24.7	11.2	0.57	312.7	9.66	13.4
Mauritania	2008	6.4	2.4	1.16	73.7	9.24	4.9
Mauritius	2008	22.1	11.0	3.92	158.3	432.84	41.4
Mozambique	2008	17.7	8.4	0.38	263.3	2.12	29.0
Namibia	2008	21.6	14.5	2.03	70.2	85.07	16.5
Niger	2008	6.2	2.6	0.17	93.4	1.53	8.3
Nigeria	2008	403.9	185.3	0.42	2,062.0	32.88	21.3
Rwanda	2008	18.1	6.1	0.53	104.2	4.54	14.6
Seychelles	2008	0.2	0.0	0.70	1.0	8.54	na
Senegal	2008	25.4	12.1	0.87	141.1	6.78	9.9
Sierra Leone	2008	5.9	2.2	0.31	66.6	4.63	5.2
South Africa	2008	272.1	165.7	2.02	783.9	215.81	40.1
Sudan	2008	51.5	22.8	0.27	1,020.5	7.47	36.2
Tanzania	2008	77.1	27.0	0.50	673.5	4.78	21.3
Togo	2008	8.7	4.0	0.47	62.7	6.25	9.9
Uganda	2008	87.7	30.5	1.24	298.7	8.33	21.5

Source: *PPP = purchasing power parity. **FTE = full-time equivalent. **Notes:** Table includes only countries where ASTI has conducted survey rounds since 2002. Public agricultural research and development (R&D) includes government, higher-education, and nonprofit agencies but excludes the private sector. Purchasing power parities (PPPs) measure the relative purchasing power of currencies across countries by eliminating national differences in pricing levels for a wide range of goods and services. PPPs are relatively stable over time, whereas exchange rates fluctuate considerably. Measuring researchers in full-time equivalents (FTEs) takes into account the proportion of time researchers spend on R&D activities. For example, four university professors who spend 25 percent of their time on research would individually represent 0.25 FTEs and collectively be counted as one FTE.

Table 1, continued

Country	Latest data available	Public research spending		Public spending as a share of ag. GDP (%)	FTE public researchers	Researchers per million economically engaged population in agriculture	Women as share of total researchers (%)
		2005 PPP dollars (millions)	2005 US dollars (millions)				
Zambia	2008	8.1	4.4	0.29	208.5	2.58	22.9
Zimbabwe	2008	na	na	na	147.7	6.83	43.2
ASIA-PACIFIC							
Bangladesh	2009	125.90	44.30	0.33	2081.30	27.59	15.8
China	2007	3,679.5	1,548.1	0.49	na	na	na
India	2003	1,426.4	474.4	0.36	16,703.7	60.87	13.9
Indonesia	2003	204.2	82.8	0.20	4,892.6	97.74	27.7
Korea	2002/03	522.7	402.6	1.66	2,201.8	na	40.0
Lao PDR	2003	10.4	2.9	0.24	123.4	56.92	22.3
Malaysia	2002	446.5	204.4	1.9	1,117.6	622.62	33.9
Myanmar	2003	4.6	4.6	0.06	618.7	32.60	54.2
Nepal	2009	23.0	7.3	0.24	388.6	33.15	9.5
Pakistan	2003	184.9	59.4	0.29	3,442.5	129.29	5.7
Papua New Guinea	2002	19.9	9.2	0.5	107.4	55.65	16.0
Philippines	2002	141.4	55.9	0.4	3,212.5	252.86	54.2
Sri Lanka	2003	52.6	18.4	0.52	577.2	138.58	32.6
Vietnam	2002/03	55.9	16.6	0.2	2,950.8	102.65	31.4
LATIN AMERICA AND THE CARIBBEAN							
Argentina	2006	448.6	196.1	1.27	3,947.3	2745.01	41.7
Belize	2006	2.6	1.3	0.95	16.7	596.43	31.1
Brazil	2006	1,306.3	728.0	1.80	5,375.5	441.59	33.8
Chile	2006	98.1	58.4	1.22	690.3	702.97	29.7
Colombia	2006	152.4	71.1	0.50	998.9	280.74	31.7
Costa Rica	2006	29.9	15.9	0.93	282.9	865.08	27.4
Dominican Republic	2006	17.4	10.3	0.26	138.8	278.67	24.9
El Salvador	2006	5.7	0.3	0.15	76.9	124.26	14.8
Guatemala	2006	8.3	4.2	0.06	102.4	53.18	14.7
Honduras	2006	12.7	4.7	0.43	123.7	181.31	7.4
Mexico	2006	517.6	338.5	1.21	4,066.7	483.38	22.3
Nicaragua	2006	24.1	8.1	0.94	133.4	364.34	26.8
Panama	2006	10.0	5.7	0.50	166.7	653.57	16.1
Paraguay	2006	3.1	9.7	0.20	128.3	163.02	32.1
Uruguay	2006	59.8	32.4	1.99	400.4	2107.58	42.5

Table 1, continued

Country	Latest data available	Public research spending		Public spending as a share of ag. GDP (%)	FTE public researchers	Researchers per million economically engaged population in agriculture	Women as share of total researchers (%)
		2005 PPP dollars (millions)	2005 US dollars (millions)				
MIDDLE EAST AND NORTH AFRICA							
Iran (Islamic Republic of)	2004	559.3	166.9	0.82	4,788.4	772.20	14.4
Jordan	2002/03	6.8	3.6	1.59	212.7	1849.13	13.2
Morocco	2002	128.6	70.8	0.95	654.8	208.54	18.4
Syrian Arab Republic	2002	77.7	136.4	0.45	1,358.1	1020.36	22.1
Tunisia	2002	51.2	22.9	0.91	440.8	574.00	27.6

SPEED DATABASE: STATISTICS ON PUBLIC EXPENDITURE FOR ECONOMIC DEVELOPMENT

The Statistics on Public Expenditure for Economic Development (SPEED) database is a resource that houses information on agricultural and other sectoral public expenditures in 70 developing countries from 1980 to 2009. (See, for example, spending details from 2005 in Table 2.) IFPRI researchers have compiled data from multiple sources, including the International Monetary Fund, the World Bank, and national governments and conducted extensive data checks and adjustments to ensure consistency over time that is free of exchange-rate fluctuations and currency-denomination changes. The SPEED database is updated periodically.

Policymakers, researchers, and other stakeholders can use this robust database for many purposes. The data allows users to examine historic trends and composition of government resource allocation across sectors and make comparisons with other countries in the region or at a similar level of development. It also aids in the analysis of alignment of actual expenditure and broad development priorities in economic growth, poverty reduction, and food security to evaluate the link between strategy priorities and budget execution, monitor progress toward achieving development goals, identify funding gaps, and diagnose the efficiency and cost effectiveness of public spending.

Website: www.ifpri.org/ourwork/programs/priorities-public-investment/speed-database
Contact: Bingxin Yu (b.yu@cgiar.org)

TABLE 2 Agricultural public expenditure for economic development, by country

Country	Agricultural expenditure, 2005 US dollars (billions)				Agricultural expenditure, 2005 international dollars (billions)				Per capita agricultural expenditure of total population, 2005 international dollars (billions)				Ratio of agricultural expenditure to agricultural GDP (%)			
	1980	1995	2009		1980	1995	2009		1980	1995	2009		1980	1995	2009	
EAST ASIA AND PACIFIC																
China	7.14	8.87	68.6		17.20	21.37	165.17		17.1	17.2	121.0		10.9	4.5	19.1	
Fiji	0.03	0.03	0.03	[1]	0.04	0.03	0.03	[1]	65.9	43.3	39.6	[1]	8.4	5.8	6.5	[1]
Indonesia	1.80	1.80	1.56	[1]	4.44	4.43	3.86	[1]	29.4	22.2	16.6	[1]	9.3	4.8	3.5	[1]
Malaysia	0.75	0.97	2.32		1.68	2.16	5.17		121.5	104.4	185.0		10.8	8.7	15.2	
Mongolia		0.01	0.05	[1]		0.02	0.15	[1]		8.8	57.8	[1]		1.4	8.8	[1]
Myanmar	19.26	12.39	18.30	[1]	0.44	0.29	0.42	[1]	13.5	6.8	9.0	[1]	8.0	2.7	1.6	[1]
Papua New Guinea	0.08	0.05	0.02	[1]	0.19	0.12	0.06	[1]	57.6	25.4	8.9	[1]	8.3	3.1	1.2	[1]
Philippines	0.43	0.86	1.14		1.08	2.16	2.89		22.8	31.2	31.5		3.2	5.7	7.2	
Thailand	0.76	2.14	2.02		1.92	5.42	5.11		40.4	90.8	74.4		7.8	16.8	9.0	
Vanuatu	0.001	0.003	0.003	[1]	0.003	0.01	0.01	[1]	23.0	31.5	25.3	[1]	4.4	5.5	3.2	[1]
Vietnam		0.33	0.71			1.12	2.38			15.1	27.4			4.6	4.7	
EUROPE AND CENTRAL ASIA																
Azerbaijan		0.08	0.19	[1]		0.27	0.61	[1]		34.4	69.6	[1]		6.2	12.5	[1]
Belarus		0.13	1.13			0.36	3.12			34.7	323.5			4.8	30.2	
Bulgaria		0.02	0.14			0.04	0.37			5.0	48.6			0.5	8.8	
Georgia			0.03				0.07				17.0				3.9	
Kazakhstan		0.11	0.64			0.25	1.47			15.4	92.9			2.6	14.3	
Kyrgyz Republic		0.02	0.02	[1]		0.05	0.06	[1]		12.0	11.2	[1]		2.2	1.9	[1]
Latvia		0.06	0.43			0.10	0.81			41.6	357.3			7.3	83.7	
Lithuania		0.28	0.34			0.51	0.62			140.8	187.0			17.7	36.6	
Moldova		0.01	0.06			0.03	0.16			7.8	43.9			1.5	16.7	
Romania	2.41	1.95	1.45	[1]	5.00	4.06	3.02	[1]	225.2	178.9	139.7	[1]	12.3	11.4	14.5	[1]
Russia		0.22	1.94			0.48	4.32			3.2	30.2			0.6	4.6	
Turkey	0.54	0.58	2.98	[1]	0.90	0.97	4.99	[1]	20.3	16.4	71.2	[1]	0.2	1.1	6.3	[1]
Ukraine			0.54				1.64				35.8				7.5	

Notes: [1] Data are from 2007. [2] Data are from 2008. [3] Data are from 2004.

Table 2, continued

	Agricultural expenditure, 2005 US dollars (billions)				Agricultural expenditure, 2005 international dollars (billions)				Per capita agricultural expenditure of total population, 2005 international dollars (billions)				Ratio of agricultural expenditure to agricultural GDP (%)			
Country	1980	1995	2009		1980	1995	2009		1980	1995	2009		1980	1995	2009	
LATIN AMERICA AND THE CARIBBEAN																
Argentina	0.00	0.14	0.11	[1]	0.00	0.32	0.25	[1]	0.003	9.2	6.4	[1]		1.7	0.6	[1]
Bolivia	0.01	0.003	0.03	[1]	0.03	0.01	0.11	[1]	5.1	1.6	11.7	[1]		0.3	2.2	[1]
Brazil		2.98	0.61	[1]		5.18	1.06	[1]		32.0	5.6	[1]		7.4	1.1	[1]
Chile	0.18	0.17	0.41	[2]	0.30	0.28	0.68	[2]	26.6	19.6	40.5	[2]	6.8	2.3	8.4	[2]
Costa Rica	0.06	0.04	0.19		0.13	0.08	0.37		53.9	22.1	79.7		4.7	2.2	10.8	
Dominican Republic	0.30	0.19	0.13	[1]	0.53	0.33	0.23	[1]	90.8	41.6	24.3	[1]	12.1	9.4	5.0	[1]
El Salvador	0.03	0.03	0.01		0.50	0.53	0.10		107.6	92.4	16.5		13.4	13.9	2.3	
Guatemala	0.14	0.05	0.06		0.26	0.09	0.12		37.0	8.9	8.7		3.8	1.3	1.6	
Jamaica		0.08	0.10			0.13	0.16			51.6	58.6			7.3	14.2	
Mexico	8.58	3.17	4.39	[1]	13.17	4.87	6.75	[1]	191.5	52.8	61.8	[1]	20.2	9.4	13.3	[1]
Panama	0.11	0.03	0.12	[1]	0.22	0.06	0.23	[1]	110.3	22.9	69.9	[1]	18.2	3.6	9.3	[1]
St. Vincent and the Grenadines	0.002	0.004	0.003	[1]	0.004	0.01	0.01	[1]	37.7	60.2	51.2	[1]	6.3	7.1	7.9	[1]
Uruguay	0.04	0.04	0.06	[1]	0.07	0.08	0.11	[1]	23.2	25.6	34.0	[1]	1.6	3.5	3.0	[1]
MIDDLE EAST AND NORTH AFRICA																
Algeria		0.53	1.23			1.21	2.84			42.8	81.4			7.2	9.6	
Egypt	0.66	0.90	0.87	[1]	2.39	3.25	3.15	[1]	53.1	52.4	40.9	[1]	12.0	8.8	5.8	[1]
Iran	0.97	1.07	0.78		3.27	3.59	2.62		84.7	60.1	35.8		7.1	4.9	3.4	
Jordan	0.02	0.11	0.05	[2]	0.03	0.19	0.08	[2]	14.3	44.5	13.7	[2]	5.0	32.4	11.0	[2]
Lebanon		0.02	0.02	[2]		0.04	0.03	[2]		11.6	8.4	[2]		1.8	1.1	[2]
Morocco	0.54	0.54	0.40	[1]	0.98	0.97	0.72	[1]	50.0	36.0	23.3	[1]	11.5	9.1	4.4	[1]
Syria	1.26	2.40	2.02		0.72	1.37	1.15		80.3	96.4	57.4		8.1	8.1	5.3	
Tunisia	0.53	0.45	0.52		1.15	0.97	1.13		177.6	108.0	109.4		28.4	17.3	15.1	
Yemen	0.02	0.04	0.07	[1]	0.05	0.10	0.20	[1]	5.7	6.9	9.1	[1]		1.7	2.4	[1]
SOUTH ASIA																
Bangladesh	0.18	0.24	0.85		0.51	0.68	2.42		6.3	5.8	16.5		2.9	2.6	6.2	
Bhutan	0.01	0.03	0.04		0.03	0.09	0.12		72.0	170.2	168.3		17.0	21.6	21.3	
India	1.82	3.50	13.07		5.47	10.51	39.21		7.8	10.9	32.5		2.5	2.9	6.4	
Maldives	0.003	0.02	0.005		0.01	0.04	0.01		33.3	141.6	25.2		0.1	0.8	0.3	
Nepal	0.06	0.09	0.14		0.20	0.28	0.43		13.3	12.8	14.5		3.8	3.8	4.1	
Pakistan	0.12	0.08	1.07		0.37	0.24	3.35		4.6	1.9	19.6		1.3	0.4	3.8	
Sri Lanka	0.16	0.24	0.40	[2]	0.46	0.70	1.14	[2]	30.8	38.2	55.7	[2]	7.4	6.7	10.0	[2]

Table 2, continued

Country	Agricultural expenditure, 2005 US dollars (billions)				Agricultural expenditure, 2005 international dollars (billions)				Per capita agricultural expenditure of total population, 2005 international dollars (billions)				Ratio of agricultural expenditure to agricultural GDP (%)			
	1980	1995	2009		1980	1995	2009		1980	1995	2009		1980	1995	2009	
SUB-SAHARAN AFRICA																
Botswana	0.05	0.12	0.10	[1]	0.11	0.25	0.22	[1]	106.8	159.8	115.8	[1]	21.6	49.1	44.2	[1]
Cape Verde		0.002	0.02			0.002	0.02			6.2	44.3			2.5	14.6	
Ethiopia	0.05	0.12	0.45	[1]	0.20	0.45	1.70	[1]	5.5	7.8	21.9	[1]	1.5	4.0	6.4	[1]
Ghana	0.06	0.02	0.01	[1]	0.13	0.04	0.03	[1]	12.0	2.3	1.3	[1]	0.1	0.6	0.4	[1]
Kenya	0.18	0.27	0.23		0.47	0.70	0.60		28.8	25.6	15.1		6.4	6.3	4.5	
Lesotho	0.01	0.07	0.03	[2]	0.03	0.12	0.06	[2]	20.5	69.3	26.3	[2]	11.3	32.4	25.5	[2]
Liberia	0.02	0.01	0.000	[2]	0.05	0.02	0.000	[2]	25.9	9.1	0.05	[2]	3.9	10.7	0.02	[2]
Malawi	0.05	0.05	0.03	[1]	0.15	0.16	0.08	[1]	24.1	16.1	6.2	[1]	8.1	8.5	2.8	[1]
Mauritius	0.03	0.05	0.07		0.07	0.10	0.14		69.6	87.7	110.5		14.2	11.7	22.2	
Namibia		0.10	0.11	[1]		0.15	0.16	[1]		92.1	72.4	[1]		17.7	13.7	[1]
Nigeria	0.53	0.14	0.23	[1]	1.16	0.31	0.51	[1]	15.3	2.8	3.5	[1]	1.8	0.4	0.5	[1]
Niger		0.03	0.04			0.07	0.08			7.6	5.5		2.9	3.0	2.3	
Seychelles		0.01	0.002	[2]		0.01	0.004	[2]		152.2	44.0	[2]		25.0	10.7	[2]
South Africa		1.13	2.07			1.84	3.38			44.4	67.9			16.2	24.6	
Swaziland	0.03	0.02	0.06	[1]	0.05	0.04	0.12	[1]	82.0	43.5	104.4	[1]	16.8	10.1	30.4	[1]
Uganda	0.003	0.01	0.08		0.01	0.03	0.23		0.7	1.3	7.1		0.6	0.4	2.2	
Zambia	0.39	0.03	0.18	[1]	0.73	0.06	0.32	[1]	126.1	7.0	26.9	[1]	55.5	3.7	10.0	[1]
Zimbabwe	0.21	0.26	0.33	[1]	0.14	0.17	0.22	[1]	19.8	14.8	18.0	[1]	13.5	10.3	12.5	[3]
HIGH-INCOME OECD COUNTRIES																
Australia	1.53	1.86	1.72		1.44	1.74	1.61		97.6	96.1	73.6		6.2	10.9	9.5	
Canada	2.61	2.16	2.99		2.63	2.18	1.55		107.3	74.4	46.1		10.7	9.1	15.1	
Czech Republic		0.49	1.99			0.82	3.34			79.5	319.6			10.2	60.6	
Estonia		0.003	0.01			0.004	0.01			2.9	7.3			0.7	1.7	
Finland	1.96	2.15	1.33		2.44	2.67	1.66		510.8	523.4	311.0		30.4	54.4	38.3	
Greece	1.31	0.98	0.07		2.33	1.75	0.13		242.0	163.9	11.2		9.3	11.2	0.8	
Hungary	3.38	1.27	2.57		5.24	1.97	3.99		489.7	191.0	398.5		22.6	20.4	70.0	
Iceland	0.24	0.19	0.22		0.16	0.12	0.14		682.3	449.2	432.9		24.6	15.9	17.5	
Ireland	1.51	0.83	0.76		1.88	1.03	0.95		549.4	286.5	474.5		37.8	18.9	58.7	
Italy	0.002	0.10	1.38		0.002	0.14	1.91		0.04	2.5	31.7		0.001	0.3	6.5	
Israel	0.71	0.61	0.22	[1]	0.86	0.74	0.27	[1]	230.9	138.6	38.4	[1]	17.8	12.9	3.7	[1]
Japan	16.09	10.52	15.90		13.68	8.95	13.52		118.0	71.9	106.8		17.3	13.6	25.8	
Korea, Rep.	1.54	8.57	11.36		2.00	11.13	14.75		53.4	249.2	307.6		5.8	24.9	41.0	
Luxembourg	0.07	0.07	0.11		0.09	0.09	0.13		255.7	217.7	263.8		49.0	47.2	135.3	
New Zealand	1.18	0.27	0.49		1.12	0.26	0.46		354.3	70.8	106.4		19.5	4.6	7.6	
Norway	3.85	3.24	2.10		2.79	2.35	1.52		682.0	538.1	314.9		64.0	46.3	50.9	
Poland		1.05	2.94			1.78	5.02			46.4	131.1			6.5	21.6	

Table 2, continued

Country	Agricultural expenditure, 2005 US dollars (billions)				Agricultural expenditure, 2005 international dollars (billions)				Per capita agricultural expenditure of total population, 2005 international dollars (billions)				Ratio of agricultural expenditure to agricultural GDP (%)			
	1980	1995	2009		1980	1995	2009		1980	1995	2009		1980	1995	2009	
Portugal		0.81	0.44			1.44	0.78			142.0	73.3			15.8	14.6	
Slovenia		0.11	0.17			0.23	0.36			117.4	176.2			16.7	27.6	
Spain	1.98	0.27	0.82		3.08	0.42	1.28		82.1	10.7	28.0		7.8	1.2	4.2	
Sweden	2.28	1.18	0.81		1.81	0.94	0.64		217.8	106.2	68.7		22.8	14.4	11.9	
Switzerland	2.29	3.12	2.52	[2]	1.69	2.30	1.85	[2]	267.4	327.3	244.4	[2]	26.1	47.3	47.6	[2]
United Kingdom	5.41	1.49	3.67		4.96	1.37	3.36		87.8	23.5	54.3		21.5	4.9	22.9	
United States	16.25	11.89	20.11		16.25	11.89	20.11		70.7	44.6	65.4		10.9	8.2	16.0	
HIGH-INCOME NON-OECD COUNTRIES																
Bahamas	0.01	0.01	0.01		0.01	0.02	0.02		55.2	66.7	50.5		12.0	10.9	13.4	
Bahrain	0.01	0.01	0.01	[2]	0.02	0.02	0.02	[2]	45.4	39.7	22.8	[2]	16.7	16.2	9.5	[2]
Croatia		0.13	0.62			0.21	0.95			43.9	214.8			6.0	24.2	
Cyprus	0.15	0.12	0.10		0.27	0.21	0.19		398.7	240.1	171.3		47.8	29.6	43.8	
Kuwait	0.01	0.12	0.13		0.02	0.18	0.19		11.0	109.6	71.7		13.0	61.6	33.0	
Malta	0.02	0.01	0.04		0.04	0.03	0.07		109.8	77.1	175.0		31.2	17.6	46.9	
Oman	0.05	0.13	0.06		0.10	0.24	0.12	[1]	81.2	109.7	47.5	[1]	28.0	20.7	9.6	[1]
Singapore	0.02	0.03	0.04	[1]	0.03	0.04	0.06		12.6	12.1	11.1		5.3	19.3	24.4	
United Arab Emirates	0.06	0.05	0.05		0.09	0.08	0.08		86.0	34.4	11.8		2.3	3.4	2.6	

GLOBAL HUNGER INDEX

Each year, IFPRI calculates the Global Hunger Index (GHI), which is designed to comprehensively measure and track hunger globally and by country and region. To reflect the multidimensional nature of hunger, the GHI combines three equally weighted indicators into one index number: (1) the percentage of people who are undernourished; (2) the percentage of children younger than age five who are underweight; and (3) the mortality rate of children younger than age five.

According to the 2011 GHI, global hunger has fallen slowly in the past two decades. (See specific country scores for 2011 in Table 3.) From 1990 to 2011, the greatest improvements in absolute scores took place in Angola, Bangladesh, Ethiopia, Mozambique, Nicaragua, Niger, and Vietnam. Hunger worsened, however, in the Burundi, Comoros, Côte d'Ivoire, the Democratic Republic of Congo, North Korea, and Swaziland.

By highlighting successes and failures in hunger reduction and providing insights into the drivers of hunger, the GHI points to the geographic areas where policy action is most needed and suggests where policy lessons can be learned.

Website: www.ifpri.org/publication/2011-global-hunger-index
Contacts: Klaus von Grebmer or Tolulope Olofinbiyi
Email: k.vongrebmer@cgiar.org or t.olofinbiyi@cgiar.org

TABLE 3 Global Hunger Index scores (various years) by 2011 country rank

Rank	Country	1990 (data from 1988–92)	1996 (data from 1994–98)	2001 (data from 1999–2003)	2011 (data from 2004–09)
1	Gabon	8.4	6.8	7.3	5.2
2	Mauritius	8.0	7.4	6.0	5.4
2	Paraguay	7.7	5.5	5.2	5.4
4	China	11.7	9.1	6.8	5.5
4	El Salvador	10.1	9.0	5.4	5.5
4	Kyrgyz Republic	—	9.1	8.7	5.5
7	Trinidad and Tobago	6.9	7.5	6.3	5.6
8	Colombia	9.1	6.8	5.8	5.7
9	Morocco	7.7	6.7	6.1	5.9
9	Peru	14.5	10.8	9.0	5.9
11	Turkmenistan	—	10.1	8.8	6.2
12	Uzbekistan	—	9.1	10.7	6.3
13	South Africa	7.0	6.5	7.4	6.4
14	Panama	9.8	9.7	9.0	7.0
15	Guyana	14.4	8.9	7.8	7.1
16	Ecuador	13.6	10.8	9.0	7.9
16	Honduras	13.4	13.2	10.1	7.9
18	Suriname	10.4	9.4	10.0	8.0
19	Thailand	15.1	11.9	9.5	8.1
20	Ghana	21.0	16.1	13.0	8.7
21	Nicaragua	22.6	17.8	12.3	9.2
22	Armenia	—	14.4	11.3	9.5
23	Dominican Republic	14.2	11.7	10.9	10.2
24	Swaziland	9.1	12.3	12.6	10.5
25	Vietnam	25.7	21.4	15.5	11.2
26	Mongolia	16.3	17.7	14.8	11.4
27	Philippines	19.9	17.5	14.1	11.5
28	Lesotho	12.7	13.9	13.8	11.9
29	Bolivia	17.0	14.6	12.5	12.2
29	Indonesia	18.5	15.5	14.3	12.2
31	Mauritania	22.7	16.9	16.9	12.7
32	Botswana	13.4	15.5	15.9	13.2
32	Congo, Rep.	23.2	24.2	16.0	13.2
34	Senegal	18.7	19.7	19.3	13.6
35	Namibia	20.3	18.7	16.3	13.8
36	Guatemala	15.1	15.8	15.1	14.0
36	Sri Lanka	20.2	17.8	14.9	14.0
38	Benin	21.5	20.2	16.9	14.7
39	Gambia, The	15.8	20.3	16.4	15.0
40	Nigeria	24.1	21.2	18.2	15.5
41	Myanmar	29.2	25.4	22.5	16.3
42	Uganda	19.0	20.4	17.7	16.7
43	Tajikistan	—	24.4	24.5	17.0
44	Burkina Faso	23.7	22.5	21.7	17.2
45	Guinea	22.4	20.3	22.4	17.3
46	Cameroon	21.9	22.4	19.4	17.7
46	Zimbabwe	18.7	22.3	21.3	17.7
48	Côte d'Ivoire	16.6	17.6	16.4	18.0
49	Malawi	29.7	27.1	22.4	18.2
50	Kenya	20.6	20.3	19.9	18.6
51	North Korea	16.1	20.3	20.1	19.0
52	Guinea-Bissau	21.7	22.3	22.8	19.5
53	Mali	27.9	26.3	23.2	19.7
54	Cambodia	31.7	31.4	26.3	19.9
54	Nepal	27.1	24.6	23.0	19.9
56	Togo	26.6	22.2	23.6	20.1
57	Lao PDR	29.0	25.2	23.6	20.2
58	Tanzania	23.1	27.4	26.0	20.5
59	Pakistan	25.7	22.0	21.9	20.7
60	Rwanda	28.5	32.7	25.2	21.0
61	Liberia	23.5	26.9	25.8	21.5
61	Sudan	29.2	24.7	25.9	21.5
63	Djibouti	30.8	25.8	25.3	22.5
63	Madagascar	24.4	24.8	24.8	22.5
65	Mozambique	35.7	31.4	28.4	22.7
66	Niger	36.2	36.2	30.8	23.0
67	India	30.4	22.9	24.1	23.7
68	Zambia	24.7	25.0	27.6	24.0

GHI Severity Scale

≥ 30.0	Extremely Alarming	5.0–9.9	Moderate
20.0–29.9	Alarming	≤ 4.9	Low
10.0–19.9	Serious		No data

Table 3, continued

Rank	Country	1990 (data from 1988–92)	1996 (data from 1994–98)	2001 (data from 1999–2003)	2011 (data from 2004–09)
69	Angola	43.0	40.7	33.4	24.2
70	Bangladesh	38.1	36.3	27.6	24.5
71	Sierra Leone	33.0	30.5	30.7	25.2
72	Yemen, Rep.	30.4	27.8	27.9	25.4
73	Comoros	22.3	27.1	30.1	26.2
74	Central African Republic	27.6	28.6	27.7	27.0
75	Timor-Leste	—	—	26.1	27.1
76	Haiti	34.0	32.3	26.0	28.2
77	Ethiopia	43.2	39.1	34.7	28.7
78	Chad	39.2	35.8	31.0	30.6
79	Eritrea	—	37.7	37.6	33.9
80	Burundi	31.4	36.3	38.5	37.9
81	Congo, Dem. Rep.	24.0	35.2	41.2	39.0

Countries with 2011 scores less than or equal to 4.9

Country	1990 (with data from 1988-92)	1996 (with data from 1994-98)	2001 (with data from 1999-2003)	2011 (with data from 2004-2009)
Albania	8.9	5.2	8.2	<5
Algeria	6.4	7.2	5.9	<5
Argentina	<5	<5	<5	<5
Azerbaijan	—	15.0	7.8	<5
Belarus	—	<5	<5	<5
Bosnia and Herzegovina		<5	<5	<5
Brazil	7.6	6.2	5.3	<5
Bulgaria	<5	<5	<5	<5
Chile	<5	<5	<5	<5
Costa Rica	<5	<5	<5	<5
Croatia	—	<5	<5	<5
Cuba	<5	6.5	<5	<5
Egypt, Arab Rep.	7.8	6.6	<5	<5
Estonia	—	<5	<5	<5
Fiji	6.1	<5	<5	<5
Georgia	—	8.9	6.0	<5
Iran, Islamic Rep.	9.4	7.5	5.0	<5
Jamaica	6.5	5.0	<5	<5
Jordan	<5	<5	<5	<5
Kazakhstan	—	<5	5.3	<5
Kuwait	9.3	<5	<5	<5
Latvia	—	<5	<5	<5
Lebanon	<5	<5	<5	<5
Libya	<5	<5	<5	<5
Lithuania	—	<5	<5	<5
Macedonia, FYR	—	<5	<5	<5
Malaysia	9.0	6.7	6.6	<5
Mexico	7.8	<5	<5	<5
Moldova	—	5.9	5.2	<5
Montenegro	—	—	—	<5
Romania	<5	<5	<5	<5
Russian Federation		<5	<5	<5
Saudi Arabia	5.8	6.1	<5	<5
Serbia	—	—	—	<5
Slovak Republic	—	<5	<5	<5
Syrian Arab Republic	7.7	5.6	5.4	<5
Tunisia	<5	<5	<5	<5
Turkey	5.7	5.2	<5	<5
Ukraine	—	<5	<5	<5
Uruguay	<5	<5	<5	<5
Venezuela, RB	6.6	7.0	6.4	<5

FOOD POLICY RESEARCH CAPACITY INDICATORS

Local capacity to conduct food policy research and analysis is essential for developing evidence-based policies and facilitating their adoption. However, development of capacity indicators in the past has been thwarted by a lack of data and the spread of policy capacity over a wide range of institutions within a country. In 2011, IFPRI started to systematically document the food policy research capacity in 25 selected developing countries, in an effort to develop indicators that eventually can measure a country's domestic capacity to design, implement, and evaluate policies in the food, agriculture, and natural resource sectors. This study defines any socioeconomic and policy-related research in the food, agriculture, and natural resource sectors as food policy research.

At the current stage, two indicators measure food policy research capacity at the country level. The first indicator records the availability of qualified human resources for food policy research per one million people living in rural areas. (See results for 2010 in Table 4.) This indicator is based on the number of full-time PhD-equivalent researchers involved in food policy research in the key ministries and academic and research organizations in a country. To calculate this number, the total number of PhD-equivalent researchers—the sum of researchers with a PhD, a master's degree (weighted as half of a PhD) and a bachelor's degree (weighted as one-quarter of a PhD) is multiplied by the average proportion of time devoted to food policy research activities.

The second indicator measures the number of international journal articles per full-time PhD-equivalent researcher for a country. This indicator is based on the number of international journal articles produced by the researchers in the surveyed organizations of a country between 2006 and 2010. To obtain this indicator, the journal articles in the Web of Science and Econ Lit databases related to socioeconomic and food policy issues were searched against the names of organizations. The total number of publications, which is the sum of the number of journal articles published by all surveyed organizations on socioeconomic and policy issues, is divided by the total number of full-time PhD-equivalent researchers in a country. This indicator reflects the quality dimension of domestic food policy research capacity.

This preliminary set of data is the beginning of a continuing initiative that ultimately aims to generate a set of food policy capacity indicators that can be easily monitored by IFPRI over time and effectively used by decisionmakers and other stakeholders for designing capacity development interventions.

Contacts: Suresh Babu or Paul Dorosh
Email: s.babu@cgiar.org or p.dorosh@cgiar.org

TABLE 4 Food policy research capacity indicators, 2010

Country	Analysts/ researchers (head count) in 2010	Full-time equivalent analysts/ researchers with PhD in 2010	International publications produced from 2006–10	Full-time equivalent analysts/ researchers with PhD per million rural population in 2010	Publications by full-time equivalent researchers with PhD (over a 5-year period)
Afghanistan	43	3.488	0	0.135	0
Bangladesh	66	31.65	25	0.296	0.79
Benin	38	11.138	12	2.17	1.077
Burundi	35	3.25	0	0.436	0
Colombia	82	14.433	12	1.252	0.831
Ethiopia	74	21.55	21	0.315	0.974
Ghana	95	12.589	35	1.064	2.78
Guatemala	45	3.965	4	0.546	1.009
Honduras	32	1.87	5	0.481	2.674
Kenya	155	30.755	43	0.976	1.398
Liberia	34	0.515	0	0.335	0
Madagascar	186	17.913	8	1.239	0.447
Malawi	50	8.875	5	0.743	0.563
Mozambique	37	5.678	15	0.394	2.642
Nepal	26	6.375	2	0.26	0.314
Niger	28	8.563	4	0.663	0.467
Nigeria	349	78.188	21	0.983	0.269
Peru	41	9.464	16	1.146	1.691
Rwanda	37	5.84	1	0.678	0.171
Senegal	71	21.513	3	3.03	0.139
South Africa	198	64.413	183	3.364	2.841
Tanzania	91	8.413	12	0.255	1.426
Uganda	34	10.89	14	0.376	1.286
Zambia	35	9.3	9	1.119	0.968
Zimbabwe	36	9.52	11	1.227	1.155

AGRICULTURAL TOTAL FACTOR PRODUCTIVITY

Total factor productivity (TFP)—or, the total amount of output relative to the total amount of inputs used to produce that output—is a key indicator of the agricultural sector's performance. In the long run, TFP is the main driver of growth in agriculture and can be affected by policies and investment. It is determined by the efficiency of resource allocation in production given a certain technology (the "efficiency" component) and the adoption of new technologies (the "technical change" component) that allow for new and more efficient ways of producing outputs. (See Table 5.) Economic policies in the past have had large negative impacts on agricultural growth in several developing countries through price distortions that resulted in increasing inefficiency and stagnated or negative growth of TFP. On the other hand, public investment in agricultural research and development has historically been a major driver of technical change in agriculture, contributing to the process of agricultural transformation in many Asian and Latin American countries.

Contact: Alejandro Nin-Pratt (a.nin-pratt@cgiar.org)

TABLE 5 Average annual growth of agricultural total factor productivity (TFP), efficiency, and technical change (percent), various years

Region/country	1981–90			1991–2000			2001–09		
	TFP	Efficiency	Technical change	TFP	Efficiency	Technical change	TFP	Efficiency	Technical change
SUB-SAHARAN AFRICA	**0.39**	**−0.03**	**0.44**	**1.99**	**1.27**	**0.71**	**2.26**	**1.60**	**0.66**
Angola	2.01	2.01	0.00	5.26	4.94	0.31	4.25	4.25	−0.00
Burkina Faso	−3.15	−4.21	1.11	1.68	1.68	0.00	8.84	8.47	0.34
Cameroon	0.54	0.54	0.00	1.73	0.71	1.01	3.36	3.03	0.33
Cote d'Ivoire	1.73	0.85	0.87	2.62	0.00	2.62	−0.05	−0.08	0.02
Ethiopia	−0.76	−0.76	0.00	0.77	0.50	0.27	3.46	3.37	0.10
Ghana	−0.32	−0.41	0.09	6.99	5.42	1.48	3.43	0.00	3.43
Guinea	−1.73	−1.73	−0.00	−0.12	−0.13	0.00	0.56	0.56	0.00
Kenya	0.87	0.81	0.06	0.34	−0.26	0.59	−0.00	−0.55	0.54
Madagascar	−0.28	−0.28	0.00	−1.22	−1.24	0.02	0.78	0.75	0.04
Malawi	0.42	0.42	0.00	4.93	3.38	1.50	−3.03	−4.80	1.86
Mali	1.11	1.11	0.00	−1.17	−1.30	0.13	5.91	5.35	0.53
Mozambique	0.10	0.10	0.00	2.04	1.37	0.67	8.27	6.03	2.11
Nigeria	2.25	2.23	0.02	3.80	1.78	1.98	1.65	0.00	1.65
Senegal	2.23	2.23	0.00	0.14	0.13	0.00	−1.32	−1.33	0.01
South Africa	1.52	−4.30	6.08	4.54	3.31	1.19	3.24	3.24	0.00
Sudan	−1.73	−1.74	0.01	2.85	2.85	0.00	2.93	2.93	0.00
Tanzania	1.57	1.55	0.02	−0.24	−0.44	0.20	6.13	6.13	−0.00
Zambia	0.11	−0.03	0.14	1.37	0.14	1.23	0.57	−0.32	0.90
Zimbabwe	0.97	0.96	0.02	1.49	1.30	0.19	−6.06	−6.67	0.65
WEST ASIA AND NORTH AFRICA	**1.85**	**−0.37**	**2.24**	**1.75**	**0.24**	**1.51**	**2.64**	**0.52**	**2.14**
Algeria	2.34	−2.34	4.80	2.76	1.57	1.17	4.00	3.95	0.05
Egypt	1.81	0.00	1.81	2.35	0.00	2.35	1.02	−0.02	1.04
Iran	−0.39	−0.42	0.04	2.58	0.90	1.67	3.14	−0.43	3.59
Jordan	3.80	3.22	0.55	1.92	0.46	1.45	5.12	1.53	3.54
Lebanon	6.11	−0.00	6.11	1.47	−0.92	2.42	2.46	0.24	2.21
Libya	2.58	1.03	1.54	4.05	2.10	1.91	1.88	−2.08	4.05
Morocco	3.04	1.37	1.65	−0.57	−1.13	0.57	4.80	4.80	0.00
Syria	−4.55	−4.97	0.45	1.91	0.48	1.42	0.06	−3.29	3.47
Tunisia	2.45	−2.23	4.79	−0.09	−1.07	0.99	1.97	1.97	0.00
Turkey	1.27	0.63	0.64	1.11	−0.01	1.12	1.97	−1.48	3.50

Source: IFPRI calculations, based on FAOSTAT data.

Table 5, continued

Region/country	1981–90			1991–2000			2001–09		
	TFP	Efficiency	Technical change	TFP	Efficiency	Technical change	TFP	Efficiency	Technical change
EAST AND SOUTH ASIA	**−0.13**	**−0.60**	**0.46**	**1.12**	**0.02**	**1.09**	**1.92**	**−0.11**	**2.02**
Bangladesh	0.11	−0.57	0.68	2.82	0.57	2.24	3.77	0.00	3.77
China	0.90	0.62	0.28	2.96	2.90	0.06	−0.49	−2.02	1.56
Indonesia	0.56	0.00	0.56	2.93	−0.00	2.93	1.41	−0.00	1.41
India	−0.24	−0.38	0.14	0.33	−0.37	0.70	−0.54	−1.09	0.55
Sri Lanka	−0.54	−0.62	0.08	0.30	0.28	0.02	−1.74	−1.88	0.14
Mongolia	0.62	0.30	0.32	10.05	6.20	3.62	0.80	−0.92	1.73
Malaysia	0.85	−0.29	1.15	−0.88	−1.18	0.31	2.10	1.65	0.44
Nepal	−0.02	−0.66	0.65	0.07	−0.37	0.44	8.76	2.32	6.30
Pakistan	0.06	−0.24	0.30	0.10	−1.17	1.28	−0.79	−2.23	1.48
Philippines	−0.83	−1.50	0.68	0.58	0.19	0.40	6.86	1.49	5.30
Thailand	−3.71	−3.81	0.10	0.06	−0.53	0.59	1.58	0.49	1.08
Vietnam	0.63	0.00	0.63	−5.86	−6.28	0.45	1.36	0.87	0.48
LATIN AMERICA	**0.54**	**−2.23**	**2.87**	**1.78**	**1.13**	**0.63**	**1.87**	**1.36**	**0.51**
Argentina	0.96	0.49	0.47	−0.36	−1.34	1.00	4.01	2.10	1.87
Bolivia	0.19	0.19	0.00	2.04	2.04	0.00	−0.05	−0.05	0.00
Brazil	3.17	−1.43	4.66	1.04	0.14	0.90	5.19	5.19	0.00
Chile	0.84	−5.29	6.47	0.94	0.15	0.79	3.65	3.65	0.00
Colombia	0.45	−3.06	3.63	3.65	3.12	0.51	2.31	2.31	0.00
Costa Rica	3.06	−3.01	6.25	2.72	2.04	0.66	3.26	3.26	0.00
Dominican Republic	0.63	−0.31	0.94	1.20	0.43	0.76	4.32	0.00	4.32
Ecuador	0.32	−1.60	1.96	−1.40	−1.71	0.32	−0.90	−0.90	−0.00
El Salvador	−1.29	−4.06	2.88	1.33	0.88	0.44	3.27	3.27	0.00
Guatemala	0.91	−0.98	1.91	1.41	0.94	0.46	1.44	0.87	0.57
Haiti	−0.77	−0.77	0.00	−1.28	−1.31	0.03	−0.71	−0.75	0.04
Mexico	−1.49	−5.40	4.14	2.34	1.22	1.11	2.32	1.95	0.36
Nicaragua	−0.22	−4.10	4.05	5.68	5.21	0.45	−2.67	−2.67	0.00
Panama	0.24	−3.50	3.87	0.87	0.09	0.78	3.31	3.31	0.00
Peru	1.40	−0.39	1.80	3.31	3.22	0.08	2.05	2.04	0.01
Uruguay	1.37	0.59	0.78	1.73	0.99	0.74	1.12	−0.35	1.47
Venezuela	−0.54	−5.20	4.91	4.99	3.18	1.75	−0.07	−0.07	0.00
ALL REGIONS	**0.66**	**−0.81**	**1.50**	**1.66**	**0.67**	**0.98**	**2.17**	**0.84**	**1.33**

NOTES

CHAPTER 1

Text

[1] N. Kumar and A. R. Quisumbing, *Gendered Impacts of the 2007–08 Food Price Crisis: Evidence Using Panel Data from Rural Ethiopia,* IFPRI Discussion Paper No. 1093 (Washington, DC: International Food Policy Research Institute, 2011).

[2] Centre for Research on the Epidemiology of Disasters (CRED), Catholic University of Louvain (Belgium), EM-DAT: The International Disaster Database, www.emdat.be/database.

[3] Munich Re, "Review of Natural Catastrophes in 2011: Earthquakes Result in Record Loss Year," press release, January 4, 2012.

[4] H. De Gorter and D. Just, "The Economics of Blend Mandates for Biofuels," *American Journal of Agricultural Economics* 91, no. 3 (2009): 738–750.

[5] Food and Agriculture Organization of the United Nations, *The State of the World's Land and Water Resources: Managing Systems at Risk* (Rome and London: FAO and Earthscan, 2011).

[6] J. Behrman, R. Meinzen-Dick, and A. Quisumbing, "The Gender Implications of Large-Scale Land Deals," *Journal of Peasant Studies* 39, no. 1 (2012): 49–79.

[7] High Level Panel of Experts on Food Security and Nutrition of the Committee on World Food Security, *Land Tenure and International Investments in Agriculture* (Rome: 2011); Food and Agriculture Organization of the United Nations, "FAO Head Warns on Land-Grabbing: Foreign Investment as Tool for Development," press release, May 12, 2011; K. Deininger and D. Byerlee with J. Lindsay, A. Norton, H. Selod, and M. Stickler, *Rising Global Interest in Farmland: Can It Yield Sustainable and Equitable Benefits?* (Washington, DC: World Bank); International Fund for Agricultural Development, *Responding to "Land Grabbing" and Promoting Responsible Investment in Agriculture*, IFAD Occasional Paper Series (Rome: 2011).

CHAPTER 2

Text

[1] C. Martins-Filho, M. Torero, and F. Yao, "Estimation of Quantiles Based on Nonlinear Models of Commodity Price Dynamics and Extreme Value Theory" (Washington, DC: International Food Policy Research Institute, 2010), mimeo.

[2] Figure 2 shows the results of a model of the dynamic evolution of daily returns based on historical data going back to 1954 (known as the Nonparametric Extreme Quantile [NEXQ] Model). This model is then combined with extreme value theory to estimate higher-order quantiles of the return series, allowing for classification of any particular realized return (that is, effective return in the futures market) as extremely high or not. A period of time characterized by extreme price variation (volatility) is a period of time in which we observe a large number of extreme positive returns. An extreme positive return is defined to be a return that exceeds a certain pre-established threshold. This threshold is taken to be a high order (95 percent) conditional quantile (that is, a value of return that is exceeded with low probability: 5 percent). One or two such returns do not necessarily indicate a period of excessive volatility. Periods of excessive volatility are identified based on a statistical test applied to the number of times the extreme value occurs in a window of 60 consecutive days.

[3] P. Al-Riffai, B. Dimaranan, and D. Laborde, "Global Trade and Environmental Impact Study of the EU Biofuels Mandate," report on a study carried out by the International Food Policy Research Institute (IFPRI) for the Directorate General on Trade of the European Commission (Brussels, 2010), mimeo; P. Al-Riffai, B. Dimaranan, and D. Laborde, "European Union and United States Biofuel Mandates: Impacts on World Markets," Technical Notes No. IDB-TN-191 (Washington, DC: Inter-American Development Bank, 2010); D. Laborde, *Assessing the Land Use Changes Consequences of European Biofuel Policies and Its Uncertainties,* ATLASS research report for the Directorate General on Trade of the European Commission (Brussels, 2011).

[4] Organisation for Economic Co-operation and Development, *Rising Food Prices: Causes and Consequences*, OECD Policy Report (Paris, 2008), www.oecd.org/dataoecd/54/42/40847088.pdf; J. von Braun, *Rising Food Prices: What Should Be Done?* IFPRI Policy Brief (Washington, DC: International Food Policy Research Institute, 2008); US Department of Agriculture, World Agricultural Outlook Board, *World Agricultural Supply and Demand Estimates,* WASDE-460 (Washington, DC, 2008); D. Headey and S. Fan, *Reflections on the Global Food Crisis: How Did It Happen? How Has It Hurt? And How Can We Prevent the Next One?* Research Monograph 165 (Washington, DC: International Food Policy Research Institute, 2010); HM Government, *The 2007/08 Agricultural Price Spikes: Causes and Policy Implications* (London: Department for Environment, Food and Rural Affairs, 2010).

[5] R. H. Kripalani, J. H. Oh, A. Kulkarni, S. S. Sabered, and H. S. Chaudhari, "South Asian Summer Monsoon Precipitation Variability: Coupled Climate Model Simulations and Projections under IPCC AR4," *Theoretical and Applied Climatology* 90, nos. 3–4 (2007): 133–159; M. J. Salinger, "Climate Variability and Change: Past, Present, and Future—An Overview," *Climatic Change* 70, nos. 1–2 (2005): 9–29; K. Viatcheslav, F. W. Zwiers, X. Zhang, and G. C. Hegerl, "Changes in Temperature and Precipitation Extremes in the IPCC Ensemble of Global Coupled Model Simulations," *Journal of Climate* 20, no. 8 (2010): 1419–1444; F. Giorgi, X. Bi , and J. Pal, "Mean, Interannual Variability and Trends in a Regional Climate Change Experiment over Europe. II: Climate Change Scenarios (2071–2100)," *Climate Dynamics* 23, nos. 7–8 (2004): 839–858.

[6] G. C. Nelson, M. W. Rosegrant, A. Palazzo, I. Gray, C. Ingersoll, R. Robertson, S. Tokgoz, et al., *Food Security, Farming, and Climate Change to 2050: Scenarios, Results, Policy Options* (Washington, DC: International Food Policy Research Institute, 2010).

[7] M. Hernandez and M. Torero, "Examining the Dynamic Relationship between Spot and Futures Prices of Agricultural Commodities," in *Commodity Market Review 2009–2010* (Rome: Food and Agriculture Organization of the United Nations, 2010).

[8] W. Martin and K. Anderson, "Trade Distortions and Food Price Surges," paper prepared for the World Bank–University of California, Berkeley, conference "Agriculture for Development—Revisited," Berkeley, CA, USA, October 1–2, 2010.

[9] B. Wright, *International Grain Reserves and Other Instruments to Address Volatility in Grain Markets,* Policy Research Working Paper 5028 (Washington, DC: World Bank, 2009).

[10] P. Timmer, "Reflections on Food Crises Past," *Food Policy* 35, no. 1 (2010): 1–11.

[11] B. Wright, *Speculators, Storage, and the Price of Rice* (Giannini Foundation of Agricultural Economics, University of California, Berkeley, 2008); Wright, *International Grain Reserves and Other Instruments to Address Volatility;* A. Evans, *The Feeding of the Nine Billion: Global Food Security for the 21st Century* (London: Chatham House, 2009).

[12] Martins-Filho, Torero, and Yao, "Estimation of Quantiles Based on Nonlinear Models of Commodity Price Dynamics."

[13] A. Sarris, *Hedging Cereal Import Price Risks and Institutions to Assure Import Supplies,* FAO Working Paper (Rome: Food and Agriculture Organization of the United Nations, 2009).

[14] Wright, *International Grain Reserves and Other Instruments to Address Volatility;* J. Lin, "Prepared Remarks Presented at the Roundtable on 'Preparing for the Next Global Food Price Crisis,'" Center for Global Development, Washington, DC, October 17, 2008.

[15] J. von Braun and M. Torero, *Physical and Virtual Global Food Reserves to Protect the Poor and Prevent Market Failure,* Policy Brief 4 (Washington, DC: International Food Policy Research Institute, 2008).

[16] Lin, "Prepared Remarks Presented at the Roundtable"; J. von Braun, J. Lin, and M. Torero, "Eliminating Drastic Food Price Spikes: A Three-Pronged Approach for Reserves," note for discussion (International Food Policy Research Institute, Washington, DC, 2009); and for rice: Timmer, "Reflections on Food Crises Past."

[17] von Braun and Torero, *Physical and Virtual Global Food Reserves;* J. von Braun and M. Torero, *Implementing the Physical and Virtual Food Reserves to Protect the Poor and Prevent Market Failure*, IFPRI Policy Brief 10 (Washington, DC: International Food Policy Research Institute, 2009); J. von Braun and M. Torero, "Exploring the Price Spike," Choices 24, no. 1 (2009), www.choicesmagazine.org/magazine/article.php?article=58, accessed March 16, 2012.

Box 1

[1] K. Anderson and S. Nelgen, "Agricultural Trade Distortions during the Global Financial Crisis," *Oxford Review of Economic Policy* 28, no. 1 (2012).

[2] W. Martin and K. Anderson, "Export Restrictions and Price Insulation during Commodity Price Booms," *American Journal of Agricultural Economics* 94, no. 2 (2012): 422–27.

CHAPTER 3

Text

[1] S. Devereux, *Vulnerable Livelihoods in Somali Region, Ethiopia,* Research Report No. 57 (Sussex, UK: Institute for Development Studies, 2006); C. Funk, *A Climate Trend Analysis of Kenya: August 2010,* US Geological Survey Fact Sheet 2010-3074 (Washington, DC: US Geological Survey, 2010); A. P. Williams and C. Funk, "A Westward Extension of the Warm Pool Leads to a Westward Extension of the Walker Circulation, Drying Eastern Africa," *Climate Dynamics* 37, nos. 11–12 (2011): 2417–2435.

[2] A. Sen, *Poverty and Famines: An Essay on Entitlement and Deprivation* (Oxford, UK: Clarendon Press, 1981).

[3] S. Devereux, "Introduction: From 'Famines' to 'New Famines,' in *The New Famines: Why Famines Persist in an Era of Globalization*, ed. S. Devereux (London: Routledge, 2007).

[4] See, for example, J. G. McPeak, P. D. Little, and C. R. Doss, *Risk and Social Change in an African Rural Economy: Livelihoods in Pastoralist Communities* (New York: Routledge, 2011); F. Flintan, *Broken Lands: Broken Lives? Causes, Processes, and Impacts of Land Fragmentation in the Rangelands of Ethiopia, Kenya, and Uganda* (Nairobi: REGLAP [Regional Learning and Advocacy Programme], 2011); and Devereux, *Vulnerable Livelihoods in Somali Region, Ethiopia.*

[5] J. F. Maystadt, A. Mabiso, and O. Ecker, "Climate Change and Civil War in Somalia: Does Drought Fuel Conflict through Livestock Price Shocks?" paper presented at the US Agency for International Development/International Food Policy Research Institute Workshop on Enhancing Resilience in the Horn of Africa, Washington, DC, December 13–14, 2011.

[6] T. J. Lybbert, C. B. Barrett, S. Desta, and D. L. Coppock, "Stochastic Wealth Dynamics and Risk Management among a Poor Population," *Economic Journal* 114, no. 498 (2004): 750–777; McPeak, Little, and Doss, *Risk and Social Change in an African Rural Economy*; Flintan, *Broken Lands*; Devereux, *Vulnerable Livelihoods in Somali Region, Ethiopia.*

[7] I. Livingstone, "The Common Property Problem and Pastoralist Economic Behavior," *Journal of Development Studies* 23, no. 1 (1986): 5–19.

[8] Lybbert et al., "Stochastic Wealth Dynamics and Risk Management among a Poor Population."

[9] P. D. Little, R. Behnke, J. McPeak, and G. Gebru, *Retrospective Assessment of Pastoral Policies in Ethiopia, 1991–2008*, Report Number 1, Pastoral Economic Growth and Development Policy Assessment, study commissioned by the UK Department for International Development (London, 2010).

[10] Lybbert et al., "Stochastic Wealth Dynamics and Risk Management among a Poor Population."

[11] Flintan, *Broken Lands.*

[12] Little et al., "Retrospective Assessment of Pastoral Policies in Ethiopia, 1991–2008"; Flintan, *Broken Lands*; McPeak, Little, and Doss, *Risk and Social Change in an African Rural Economy.*

[13] Devereux, *Vulnerable Livelihoods in Somali Region, Ethiopia.*

[14] For a review, see D. Headey, A. S. Taffesse, and L. You, *Enhancing Resilience in the Horn of Africa: An Exploration into Alternative Investment Options*, IFPRI Discussion Paper 1176 (Washington, DC: International Food Policy Research Institute, 2012).

[15] Devereux, *Vulnerable Livelihoods in Somali Region, Ethiopia.*

[16] Headey, Taffesse, and You, *Enhancing Resilience in the Horn of Africa.*

[17] P. D. Little, *Somalia: Economy without State* (Bloomington, IN, USA: Indiana University Press, 2003).

[18] McPeak, Little, and Doss, *Risk and Social Change in an African Rural Economy*; Devereux, *Vulnerable Livelihoods in Somali Region, Ethiopia*.

[19] Ibid.

[20] Ibid.

[21] Little, *Somalia: Economy without State*.

[22] Headey, Taffesse, and You, *Enhancing Resilience in the Horn of Africa*.

[23] Little et al., "Retrospective Assessment of Pastoral Policies in Ethiopia, 1991–2008."

[24] Y. Aklilu and A. Catley, *Livestock Exports from the Horn of Africa: An Analysis of Benefits by Pastoralist Wealth Group and Policy Implications* (Medford, MA: Feinstein International Center, Tufts University, 2010).

[25] McPeak, Little, and Doss, *Risk and Social Change in an African Rural Economy*.

[26] Devereux, *Vulnerable Livelihoods in Somali Region, Ethiopia*.

[27] Personal communication with John Graham, USAID Ethiopia.

[28] Headey, Taffesse, and You, *Enhancing Resilience in the Horn of Africa*.

[29] J. Konyndyk, "Exploring the Links between Peacebuilding and Drought Resiliency in the Horn: Lessons from Mercy Corps' Programs," presentation at "Enhancing Resilience in the Horn of Africa," a workshop hosted by the US Agency for International Development and the International Food Policy Research Institute, Washington, DC, December 13–1.

Box 3

[1] P. A. Dorosh, S. J. Malik, and M. Krausova, "Rehabilitating Agriculture and Promoting Food Security Following the 2010 Pakistan Floods," *Pakistan Development Review* 49, no. 3 (2010): 167–192.

[2] For more information on the 2011 floods in Pakistan, see the Food and Agriculture Organization of the United Nations, "Pakistan Floods 2011," www.fao.org/emergencies/country_information/list/asia/pakistanfloods2011, accessed March 15, 2012.

Box 4

[1] For more information on the crisis in the Horn of Africa, see World Food Programme, "Horn of Africa Crisis," www.wfp.org/crisis/horn-of-africa, accessed March 15, 2012.

CHAPTER 4

Text

[1] See for example D. B. Lobell, W. Schlenker, and J. Costa-Roberts, "Climate Trends and Global Crop Production since 1980," *Science* 333, no. 6042 (2011): 616–620, http://dx.doi.org/10.1126/science.1204531.

[2] P. Ericksen, P. Thornton, A. Notenbaert, L. Cramer, P. Jones, and M. Herrero, *Mapping Hotspots of Climate Change and Food Insecurity in the Global Tropics,* CCAFS Report No. 5 (Copenhagen: CGIAR Research Program on Climate Change, Agriculture and Food Security [CCAFS], 2011), http://ccafs.cgiar.org/sites/default/files/assets/docs/ccafsreport5-climate_hotspots_final.pdf.

[3] K. K. Garg, L. Karlberg, J. Barron, S. P. Wani, and J. Rockstrom, "Assessing Impacts of Agricultural Water Interventions in the Kothapally Watershed, Southern India," *Hydrological Processes* 26, no. 3 (2011): 287–404, http://dx.doi.org/10.1002/hyp.8138.

[4] J. Wang, J. Huang, and S. Rozelle, *Climate Change and China's Agricultural Sector: An Overview of Impacts, Adaptation, and Mitigation,* Issue Brief No. 5 (Geneva and Washington, DC: International Centre for Trade and Sustainable Development and International Food and Agricultural Trade Policy Council, 2010).

[5] E. Bryan, C. Ringler, B. Okoba, J. Koo, M. Herrero, and S. Silvestri, *Agricultural Management for Climate Change Adaptation, Greenhouse Gas Mitigation, and Agricultural Productivity: Insights from Kenya,* IFPRI Discussion Paper 1098 (Washington, DC: International Food Policy Research Institute, 2011).

[6] W. Schlenker and M. J. Roberts, "Nonlinear Temperature Effects Indicate Severe Damages to US Crop Yields under Climate Change," *Proceedings of the National Academy of Sciences of the United States of America* 106, no. 37 (2009): 15594–15598, http://dx.doi.org/10.1073/pnas.0906865106.

[7] Lobell, Schlenker, and Costa-Roberts, "Climate Trends and Global Crop Production since 1980."

Box 5

[1] S. S. Yadav, R. Redden, J. L. Hatfield, H. Lotze-Campen, and A. Hall, eds., *Crop Adaptation to Climate Change* (West Sussex, England, UK: Wiley, 2011).

[2] P. Ericksen, P. Thornton, A. Notenbaert, L. Cramer, P. Jones, M. Herrero. 2011. *Mapping Hotspots of Climate Change and Food Insecurity in the Global Tropics.* CCAFS Report No. 5. CGIAR Research Program on Climate Change, Agriculture and Food Security (CCAFS). Copenhagen, Denmark. Available online at: www.ccafs.cgiar.org. http://ccafs.cgiar.org/resources/climate_hotspots

[3] CGIAR Research Program on Climate Change, Agriculture and Food Security, "Climate Change Emissions and Livelihoods at the Forest-Farm Frontier," http://ccafs.cgiar.org/news/pro-poor-mitigation/climate-change-emissions-and-livelihoods-forest-farm-frontier, accessed February 15, 2012.

[4] CGIAR Research Program on Climate Change, Agriculture and Food Security, "Meet the Gender, Climate Change, Agriculture, and Food Security Grant Recipients," http://ccafs.cgiar.org/news/decision-support/meet-gender-climate-change-agriculture-and-food-security-grant-recipients, accessed February 15, 2012.

[5] http://ccafs.cgiar.org/news/research-highlights/change-and-innovation-climate-adaptation-practices

CHAPTER 5

Text

[1] Food and Agriculture Organization of the United Nations and Organisation for Economic Co-operation and Development, *Price Volatility in Food and Agricultural Markets: Policy Responses*, policy

report including contributions by FAO, IFAD, IMF, OECD, UNCTAD, WFP, the World Bank, the WTO, IFPRI and UN HLTF (Rome and Paris, 2011), www.oecd.org/dataoecd/40/34/48152638.pdf, accessed March 19, 2012.

[2] FAO and OECD, "Price Volatility in Food and Agricultural Markets."

[3] C. Valdes, "Brazil's Ethanol Industry: Looking Forward," BIO-02 (Washington, DC: Economic Research Service, US Department of Agriculture, 2011), www.ers.usda.gov/Publications/BIO02/BIO02.pdf, accessed March 19, 2012.

[4] H. Qiu, J. Huang, J. Yang, S. Rozelle, and Y. Zhang, "Bioethanol Development in China and the Potential Impacts on Its Agricultural Economy," *Applied Energy* 87, no. 1 (2010): 76–83.

[5] A. Elobeid and C Hart, "Ethanol Expansion in the Food versus Fuel Debate: How Will Developing Countries Fare?" *Journal of Agricultural and Food Industrial Organization* 5, no. 2 (2007); M. W. Rosegrant, T. Zhu, S. Msangi, and T. Sulser, "Global Scenarios for Biofuels: Impacts and Implications," *Review of Agricultural Economics* 30, no. 3 (2008): 495–505.

[6] T. Searchinger, R. Heimlich, R. A. Houghton, F. Dong, A. Elobeid, J. Fabiosa, S. Tokgoz, D. Hayes, and T.-H. Yu, "Use of U.S. Croplands for Biofuels Increases Greenhouse Gases through Emissions from Land-Use Change," *Science* 319, no. 5867 (2008): 1238–1240; J. Fargione, J. Hill, D. Tilman, S. Polasky, and P. Hawthorne, "Land Clearing and the Biofuel Carbon Debt," *Science* 319, no. 5867 (2008): 1235.

[7] P. Al-Riffai, B. Dimaranan, and D. Laborde, *Global Trade and Environmental Impact Study of the EU Biofuels Mandate*, Final Report for the Directorate General for Trade of the European Commission (International Food Policy Research Institute, Washington, DC, 2010).

[8] D. Laborde, *Assessing the Land Use Change Consequences of European Biofuel Policies*, Final Report for the Directorate General for Trade of the European Commission (International Food Policy Research Institute, Washington, DC, 2011).

[9] H. De Gorter and D Just, "The Economics of Blend Mandates for Biofuels," *American Journal of Agricultural Economics* 91, no. 3 (2009): 738–750.

[10] J. von Braun, *The World Food Situation: New Driving Forces and Required Actions*, IFPRI Food Policy Report (Washington, DC: International Food Policy Research Institute, 2008); C. F. Runge and B. Senauer, "How Biofuels Could Starve the Poor," *Foreign Affairs* 86, no. 3 (2007): 41–53; C. F. Runge and B. Senauer, "How Ethanol Fuels the Food Crisis," *Foreign Affairs,* May 28, 2008; B. D. Wright, "The Economics of Grain Price Volatility," *Applied Economic Perspectives and Policy* 33, no. 1 (2011): 32–58.

[11] B. Babcock, *The Impact of US Biofuel Policies on Agricultural Price Levels and Volatility*, ICTSD Issue Paper No. 35 (Geneva: International Center for Trade and Sustainable Development, 2011).

[12] D. Sperling and S. Yeh, "Low Carbon Fuel Standards," *Issues in Science and Technology* (Winter 2009): 57–66; D. Sperling and S Yeh, "Toward a Global Low Carbon Fuel Standard, *Transport Policy* 17, no. 1 (2010): 47–49.

[13] M. Khanna, X. Chen, H. Huang, and H. Onal, "Land Use and Greenhouse Gas Mitigation Effects of Biofuel Policies," *University of Illinois Law Review* 2 (2011): 549–588; S. Kim and B. E. Dale, "Indirect Land Use Change for Biofuels: Testing Predictions and Improving Analytical Methodologies," *Biomass and Bioenergy* 91, no. 3 (2011): 1–6.

[14] S. Msangi, M. Batka, J. Witcover, and S. Yeh, *Global Land Use Change from US Biofuels: Finding Effective Mitigation Strategies*, National Low Carbon Fuel Standard Study discussion paper (Davis, CA, US: Institute for Transportation Studies, 2012).

[15] D. C. Headey and S. Fan, *Reflections on the Global Food Crisis: How Did It Happen? How Did It Hurt? And How Can We Prevent the Next One?* Research Monograph 165 (Washington, DC: International Food Policy Research Institute, 2010.

CHAPTER 6

Text

[1] International Food Policy Research Institute, *Agriculture, Nutrition, and Health: Exploiting the Linkages*, IFPRI 2020 Conference Brochure (Washington, DC, 2010).

[2] United Nations System, Standing Committee on Nutrition, SUN Road Map Implementation, http://www.unscn.org/en/scaling_up_nutrition_sun/.

[3] International Food Policy Research Institute, *Leveraging Agriculture for Improving Nutrition and Health: Highlights from an International Conference* (Washington, DC, 2011).

[4] S. Fan, R. Pandya-Lorch, and H. Fritschel, "Overview," in *Reshaping Agriculture for Nutrition and Health*, edited by S. Fan and R. Pandya-Lorch (Washington, D.C.: International Food Policy Research Institute, 2012).

[5] Ibid.; C. Hawkes and M. Ruel, *Value Chains for Nutrition,* IFPRI 2020 Conference Brief 4 (Washington, DC: International Food Policy Research Institute, 2011).

[6] IFPRI, *Leveraging Agriculture for Improving Nutrition and Health: Highlights from an International Conference.*

[7] IFPRI, *Leveraging Agriculture for Improving Nutrition and Health: The Way Forward* (Washington, DC, 2011); IFPRI, *Leveraging Agriculture for Improving Nutrition and Health: Powerpoint at a Glance* (Washington, DC, 2011).

Box 6

[1] For more information on the CGIAR Research Program on Agriculture for Improved Nutrition and Health, see International Food Policy Research Institute, "Agriculture for Improved Nutrition and Health (CRP4)," www.ifpri.org/ourwork/division/agriculture-improved-nutrition-and-health-crp4.

Box 7

[1] The information in this box is based on R. Paarlberg, *2020 Conference "Leveraging Agriculture for Improving Nutrition and Health" New Delhi, India, February 10–12, 2011: Interim Report on Short-Term Impact* (Washington, DC: International Food Policy Research Institute, 2011).

Box 9

[1] The Water, Energy, & Food Security NEXUS Resource Platform, "Bonn2011 Conference," www.water-energy-food.org/en/whats_the_nexus/bonn_nexus_conference.html, accessed March 15, 2012.

[2] For an overview of IFPRI's data on the food-water-energy nexus, see Veolia Water, *Finding the Blue Path for a Sustainable Economy*, White Paper, www.veoliawaterna.com/north-america-water/ressources/documents/1/19979,IFPRI-White-Paper.pdf, accessed March 15, 2012.

CHAPTER 7

Text

[1] United Nations Convention to Combat Desertification, "Land and Soil in the Context of a Green Economy for Sustainable Development, Food Security, and Poverty Eradication," The Submission of the UNCCD Secretariat to the Preparatory Process for the Rio+ 20 Conference, 2011, www.unccd.int/knowledge/menu.php.

[2] E. Nkonya, N. Gerber, P. Baumgartner, J. von Braun, A. De Pinto, V. Graw, E. Kato, J. Kloos, and T. Walter, *The Economics of Land Degradation: Toward an Integrated Global Assessment,* Development Economics and Policy Series, Vol. 66, edited by F. Heidhues, J. von Braun, and M. Zeller (Frankfurt: Peter Lang, 2011).

[3] E. Nkonya, F. Place, J. Pender, M. Mwanjololo, A. Okhimamhe, E. Kato, S. Crespo, J. Ndjeunga, and S. Traore, *Climate Risk Management through Sustainable Land Management in Sub-Saharan Africa*, IFPRI Discussion Paper 1126 (Washington, DC: International Food Policy Research Institute, 2011).

[4] N. Linacre, A. Kossoy, and P. Ambrosi, *State and Trends of the Carbon Market* (Washington, DC: World Bank, 2011).

[5] D. Celentano, E. Sills, M. Sales, and A. Veríssimo, "Welfare Outcomes and the Advance of the Deforestation Frontier in the Brazilian Amazon," *World Development* 40, no. 4 (2012): 850–64, http://dx.doi.org/10.1016/j.worlddev.2011.09.002.

[6] J. C. Cole and D. M. Liverman, "Brazil's Clean Development Mechanism Governance in the Context of Brazil's Historical Environment–Development Discourses," Carbon Management 2, no. 2 (2011): 145–60.

[7] Convention on Biological Diversity, *Global Biodiversity Outlook 3* (Montreal, 2010).

[8] Food and Agriculture Organization of the United Nations, FAOSTAT database, http://faostat.fao.org.

[9] J. Bruinsma, "The Resource Outlook to 2050," in *Proceedings of the Expert Meeting on How to Feed the World in 2050* (Rome: Food and Agriculture Organization of the United Nations, 2009).

[10] R. Licker, M. Johnston, J. A. Foley, C. Barford, C. J. Kucharik, C. Monfreda, and N. Ramankutty, "Mind the Gap: How Do Climate and Agricultural Management Explain the 'Yield Gap' of Croplands around the World?" *Global Ecology and Biogeography* 119, no. 6 (2010): 769–782.

[11] D. S. Powlson, P. J. Gregory, W. R. Whalley, J. N. Quinton, D. W. Hopkins, A. P. Whitmore, P. R. Hirsch, and K. W. T. Goulding, "Soil Management in Relation to Sustainable Agriculture and Ecosystem Services," *Food Policy* 36, no. 1 (2011): S72–S87.

[12] F. Akinnifesi, O. C. Ajayi, G. Sileshi, P. W. Chirwa, and J. Chianu, "Fertiliser Trees for Sustainable Food Security in the Maize-based Production Systems of East and Southern Africa: A Review," *Agronomy for Sustainable Development* 30, no. 3 (2010): 615–29.

[13] Nkonya et al., *Climate Risk Management.*

[14] E. Nkonya, "*N* in Africa's Country and Regional Policies: Lessons to Enhance Broader Stakeholder and Policy-maker engagement," in M. Bekunda, N. Karanja, and A. Langyintuo, eds., *Impact of Nitrogen on African Agriculture and the Environment,* International Nitrogen Initiative (Dordrecht, the Netherlands: Springer, forthcoming).

[15] B. Banful, E. Nkonya, and V. Oboh, *Constraints to Fertilizer Use in Nigeria: Insights from Agricultural Extension Service*, IFPRI Discussion Paper 1010 (Washington, DC: International Food Policy Research Institute, 2010); E. Nkonya, A. Nana, B. Bashaasha, E. Kato, and M. Magheni, "Pluralist and Demand-driven and Traditional Supply-driven Agricultural Extension Services in Africa: Which Reaches More Farmers and Women? The case of Uganda," paper presented at the Tropentag Conference "Development in the Margin," Bonn, October 5–7, 2011.

Box 10

[1] R. Kaur, "Return of Village Land," *Down to Earth,* Feb. 28, 2011, www.downtoearth.org.in/node/33020.

[2] Foundation for Ecological Security, "A Commons Story: In the Rain Shadow of the Green Revolution," Draft report presentation at the 13th International Association for the Study of the Commons Conference, Hyderabad, India, January 10–14, 2010.

Box 11

[1] Food and Agriculture Organization of the United Nations, *The State of Food and Agriculture 2010–11* (Rome, 2011), www.fao.org/publications/sofa.

[2] S. Lastarria-Cornhiel, J. A. Behrman, R. Meinzen-Dick, and A. R. Quisumbing, "Gender Equity and Land: Toward Secure and Effective Access for Rural Women," in *Gender in Agriculture and Food Security: Closing the Knowledge Gap*, eds. A. Quisumbing, R. Meinzen-Dick, T. Raney, A. Croppenstedt, J. A. Behrman, and A. Peterman (New York: Springer and FAO, forthcoming).

[3] World Bank, *World Development Report 2012: Gender Equality and Development* (Washington, DC: World Bank, 2011).

[4] Lastarria-Cornhiel et al., "Gender Equity and Land."

[5] A. Bomuhangi, C. Doss, and R. Meinzen-Dick, *Who Owns the Land? Perspectives from Rural Ugandans and Implications for Land Acquisitions*, IFPRI Discussion Paper 1136 (Washington, DC: International Food Policy Research Institute, 2011).

[6] N. Kumar and A. R. Quisumbing, "Policy Reform toward Gender Equality in Ethiopia: Little by Little the Egg Begins to Walk," unpublished manuscript (Washington, DC: International Food Policy Research Institute, 2010).

CHAPTER 8

Text

[1] J. P. Pradhan and G. Lazaroiu, "Rise of Emerging Economies: An Introduction," *Economics, Management, and Financial Markets* 6, No. 1 (2011): 8–18.

[2] Joint Declaration of the Second Meeting of BRICS Ministers of Agriculture and Agrarian Development, October 30, 2011, Chengdu, China.

[3] S. Fan and J. Brzeska, *The Role of Emerging Countries in Global Food Security*, IFPRI Policy Brief 15 (Washington, DC: International Food Policy Research Institute, 2010).

[4] Busan Partnership for Effective Development Co-operation, Fourth High-Level Forum on Aid Effectiveness, Busan, Repubic of Korea, November 29–December 1, 2011.

[5] T. Elhaut and N. Berghinz, "South-South Cooperation in IFAD's Operating Model: A Background Paper on Experiences, Lessons Learnt, and Ways Forward" (International Fund for Agricultural Development, Rome, 2010).

[6] Economic and Social Council of the United Nations, "Background Study for the Development Cooperation Forum: Trends in South-South and Triangular Development Cooperation" (New York, 2008), www.un.org/en/ecosoc/docs/pdfs/south-south_cooperation.pdf.

[7] D. Brautigam, *The Dragon's Gift: The Real Story of China in Africa*, (Oxford, UK: Oxford University Press, 2009).

[8] World Economic Forum, *Realizing a New Vision for Agriculture: A Roadmap for Stakeholders* (Geneva, 2010).

[9] World Economic Forum, Forum G20 Task Force Groups Recommendations for Food Security, news release, June 8, 2011, www.weforum.org/news/forum-g20-task-force-group-presents-recommendations-food-security.

[10] Sir Ratan Tata Trust & Navajbai Ratan Tata Trust, *Annual Report 2010–2011* (Mumbai, 2011).

[11] L.-E. Eleftheriou-Smith, "Unilever Launches Global Charity Foundation," *Marketing*, January 27, 2012, www.marketingmagazine.co.uk/news/1114363/Unilever-launches-global-charity-foundation; Publicity Update, "Unilever Launches Global Foundation," January 30, 2012, www.publicityupdate.co.za/?idstory=44564.

[12] Rockefeller Foundation, Grants and Grantees: Alliance for a Green Revolution in Africa, www.rockefellerfoundation.org/grants/grants-and-grantees/23e190b2-9cf7-4706-bc90-f4e70d036633.

[13] G20 Agriculture Ministers, Action Plan on Food Price Volatility and Agriculture, Ministerial Declaration, Meeting of G20 Agriculture Ministers, Paris, June 22–23, 2011.

[14] United Nations Conference on Trade and Development, "The Role of South-South and Triangular Cooperation for Sustainable Agriculture Development and Food Security in Developing Countries," note for the Second Session, December 14–16, 2009.

[15] Bill & Melinda Gates Foundation website, www.gatesfoundation.org, accessed December 30, 2011.

[16] S. Fan, *Halving Hunger: Meeting the First Millennium Development Goal through "Business as Unusual,"* IFPRI Food Policy Report (Washington, DC: International Food Policy Research Institute, 2010).

[17] World Economic Forum, Agriculture and Food Security, http://www.weforum.org/issues/agriculture-and-food-security?fo=1.

[18] International AgriFood Network, "Proposal for Modalities for Private Sector Participation in the Committee for World Food Security," Thirty-Seventh Session, Committee on World Food Security, Rome, October 17–22, 2011.

Box 12

[1] Conclusions in Box 12 are based on a triangulation of media reports, interviews, field visits, and published research. For further details and more specific source information, see D. Brautigam and T. Xiaoyang, "China's Engagement in African Agriculture: Down to the Countryside," *China Quarterly* 199 (2009): 686–706; D. Brautigam, *The Dragon's Gift: The Real Story of China in Africa* (Oxford: Oxford University Press, 2009); and *China in Africa: The Real Story Blog,* http://chinaafricarealstory.com.

Box 14

[1] Hudson Institute Center for Global Prosperity, *Index of Global Philanthropy and Remittances 2011* (Washington, DC: 2011).

Box 15

[1] World Economic Forum, *Realizing a New Vision for Agriculture: A Roadmap for Stakeholders* (Geneva: World Economic Forum, 2010), www.weforum.org/reports/realizing-new-vision-agriculture-roadmap-stakeholders.

[2] PepsiCo, "PepsiCo to Partner with China's Ministry of Agriculture to Promote Sustainable Farming," press release, September 19, 2011, www.pepsico.com/PressRelease/PepsiCo-to-Partner-with-Chinas-Ministry-of-Agriculture-to-Promote-Sustainable-Fa09192011.html.

[3] PepsiCo, Greater China, *Performance with Purpose: Sustainability Report*, 2011, www.pepsico.com/Download/GCR_Sustainability_Report_EN_Final.pdf.

[4] PepsiCo, India, "Purpose: Human Sustainability," http://pepsicoindia.co.in/purpose/human-sustainability.html, accessed February 13, 2012.

CHAPTER 9

Text

[1] Economist Intelligence Unit, Country Reports: Algeria, Bahrain, Comoros, Djibouti, Egypt, Iran, Iraq, Jordan, Kuwait, Lebanon, Libya, Malta, Mauritania, Morocco, Oman, Qatar, Saudi Arabia, Syria, Tunisia, United Arab Emirates, and Yemen for January, February, March, and April 2011, www.eiu.com; International Monetary Fund, *Regional Economic Outlook: Middle East and Central Asia* (Washington, DC, 2011), www.imf.org/external/pubs/ft/reo/2011/mcd/eng/mreo0411.htm.

[2] *Regional Economic Outlook: Middle East and Central Asia*; *The Economist*, "Throwing Money at the Street," March 10, 2011, www.economist.com/node/18332638?story_id=18332638.

[3] United Nations Development Programme, *Arab Human Development Report 2009: Challenges to Human Security in the Arab Countries* (New York, 2009); J. R. Harrigan and H. El-Said, "The Economic Impact of IMF and World Bank Programs in the Middle East and North Africa: A Case Study of Jordan, Egypt, Morocco, and Tunisia, 1983–2004," *Review of Middle East Economics and Finance* 6, no. 2 (2010): 1–25, www.bepress.com/rmeef/vol6/iss2/art1.; C. Breisinger, O. Ecker, and P. Al-Riffai, *Economics of the Arab Awakening: From Revolution to Transformation and Food Security*, IFPRI Policy Brief 18 (Washington, DC: International Food Policy Research Institute, 2011).

[4] International Monetary Fund, "Global Recovery Stalls, Downside Risks Intensify," *World Economic Outlook Update*, January 24 (Washington, DC: 2012).

[5] O. Badiane, *Agriculture and Structural Transformation in Africa*, Stanford Symposium Series on Global Food Policy and Food Security in the 21st Century (Stanford, CA: Center on Food Security and the Environment. Stanford University, 2011); M. McMillan and D. Rodrik, *Globalization, Structural Change, and Productivity Growth*, Working Paper 17143 (Cambridge, MA: National Bureau for Economic Research, 2011).

[6] Badiane, *Agriculture and Structural Transformation in Africa.*

[7] R. Hausmann and B. Klinger, *Structural Transformation and Patterns of Comparative Advantage in the Product Space*, Faculty Research Working Paper Series (Cambridge, MA: Harvard University, Kennedy School of Government, 2006).

[8] H. Chenery, "Patterns of Industrial Growth," *American Economic Review* 50 (1960): 624–654.

[9] D. J. Teece, "Firm Capabilities and Economic Development: Implications for Newly Industrializing Economies," in L. Kim and R. Nelson, eds., *Technology Learning and Innovation: The Experience of Newly Industrializing Economies* (Cambridge, UK: Cambridge University Press, 2000).

[10] New Partnership for Africa's Development, Comprehensive Africa Agriculture Development Plan, http://www.nepad-caadp.net/.

[11] S. Lall, "Technological Change and Industrialization in the Newly Industrializing Asian Economies: Achievements and Challenges," in L. Kim and R. Nelson, eds., *Technology Learning and Innovation: The Experience of Newly Industrializing Economies* (Cambridge, UK: Cambridge University Press, 2000); Teece, "Firm Capabilities and Economic Development."

[12] M. Dodgson, "Policies for Science, Technology, and Innovation in Asian Newly Industrializing Economies," in L. Kim and R. Nelson, eds., *Technology Learning and Innovation: The Experience of Newly Industrializing Economies* (Cambridge, UK: Cambridge University Press, 2000); J. Mathews, "High Technology Industrialization in East Asia," *Journal of Industry Studies* 3, no. 2 (1996): 1–77.

[13] New Partnership for Africa's Development, *CAADP Framework for the Improvement of Rural Infrastructure and Trade-related Capacities for Market Access: Strategic Area A: Raising Competitiveness and Seizing Opportunities in Domestic, Regional, and International Markets* (Johannesburg, 2008).

[14] T. Sonobe and K. Otsuka, *Cluster-Based Industrial Development: A Comparative Study of Asia and Africa* (New York: Palgrave Macmillan, 2011).

[15] India, Planning Commission, *Report of the Working Group on Nutrition for the Twelfth Five-Year Plan (2012–17)* (New Delhi, 2011).

[16] United Nations Development Programme, *Human Development Report 2011: Sustainability and Equity: A Better Future for All* (New York, 2011).

Box 16

[1] Regional Strategic Analysis and Knowledge Support System (ReSAKSS), www.resakss.org, accessed March 16, 2012.

[2] Ministry of Finance and Economic Planning (MINECOFIN), Republic of Rwanda, Budget Laws for 2006–2011, www.minecofin.gov.rw/ministry/directorates/nb, accessed March 26, 2012.

[3] O. Badiane, *Sustaining and Accelerating Africa's Agricultural Growth Recovery in the Context of Changing Global Food Prices*, IFPRI Policy Brief 9 (Washington, DC: International Food Policy Research Institute, 2008).

Box 17

[1] Ministry of Women and Child Development, Government of India, *Five Year Strategic Plan (2011–2016)*, http://wcd.nic.in/MWCD_Strategic_Plan_10-02-2011.pdf, accessed March 16, 2012.

[2] A. Kumar, P. Kumar, Praduman, and A. N. Sharma, "Rural Poverty and Agricultural Growth in India: Implications for the Twelfth Five Year Plan," *Indian Journal of Agricultural Economics* 66, no. 3 (2011): 269–78.

[3] S. Kadiyala, P. K. Joshi, S. Mahendra Dev, T. Nanda Kumar, and V. S. Vyas, *Strengthening the Role of Agriculture for a Nutrition-Secure India* (New Delhi, India: International Food Policy Research Institute, 2011).

CONTRIBUTORS

Perrihan Al-Riffai is a research analyst in the Development Strategy and Governance Division of the International Food Policy Research Institute, Washington, DC.

Kym Anderson is George Gollin Professor of Economics at the University of Adelaide, Australia.

Suresh Babu is a senior research fellow in the Partnerships, Impact, and Capacity Strengthening Unit of the International Food Policy Research Institute, Washington, DC.

Ousmane Badiane is director in Africa for the International Food Policy Research Institute and head of its West and Central Africa Office, Dakar, Senegal.

Nienke Beintema is program head of the Agricultural Science & Technology Indicators Initiative at the International Food Policy Research Institute, Rome.

Samuel Benin is a research fellow in the Development Strategy and Governance Division of the International Food Policy Research Institute, Kampala, Uganda.

Deborah Brautigam is a senior research fellow in the Development Strategy and Governance Division of the International Food Policy Research Institute, Washington, DC.

Clemens Breisinger is a research fellow in the Development Strategy and Governance Division of the International Food Policy Research Institute, Washington, DC.

Joanna Brzeska is a consultant for the International Food Policy Research Institute. She is based in Luxembourg.

Bruce Campbell is director of the CGIAR Research Program on Climate Change, Agriculture and Food Security, Copenhagen, Denmark.

Rahul Chaturvedi is a senior program officer at the Foundation for Ecological Security, Anand, India.

Kevin Chen is a senior research fellow in the Development Strategy and Governance Division of the International Food Policy Research Institute, Beijing.

Cindy Cox is a technical writer in the Environment and Production Technology Division of the International Food Policy Research Institute, Washington, DC.

S. Mahendra Dev is director of the Indira Gandhi Institute of Development Research, Mumbai, India.

Betina Dimaranan is a research fellow in the Markets, Trade, and Institution Division of the International Food Policy Research Institute, Washington, DC.

Paul Dorosh is director of the Development Strategy and Governance Division at the International Food Policy Research Institute, Washington, DC.

Olivier Ecker is a postdoctoral fellow in the Development Strategy and Governance Division of the International Food Policy Research Institute, Washington, DC.

Shenggen Fan is director general of the International Food Policy Research Institute, Washington, DC.

Heidi Fritschel is a senior editor at the International Food Policy Research Institute, Washington, DC.

Sara Gustafson is a communications specialist at the International Food Policy Research Institute, Washington, DC.

Derek Headey is a research fellow in the Development Strategy and Governance Division of the International Food Policy Research Institute, Addis Ababa, Ethiopia.

Jikun Huang is director of the Center for Chinese Agricultural Policy, Chinese Academy of Sciences, Beijing.

Kabba T. Joiner is a senior policy adviser at Helen Keller International, Accra, Ghana.

P. K. Joshi is director in South Asia for the International Food Policy Research Institute, New Delhi.

Suneetha Kadiyala is a research fellow in the Poverty, Health, and Nutrition Division of the International Food Policy Research Institute, New Delhi.

Zhenya Karelina was a research assistant in the Director General's Office of the International Food Policy Research Institute, Washington, DC.

Jawoo Koo is a research fellow in the Environment and Production Technology Division of the International Food Policy Research Institute, Washington, DC.

David Laborde is a senior research fellow in the Markets, Trade, and Institutions Division of the International Food Policy Research Institute, Washington, DC.

Rachel Licker is a graduate research assistant at the University of Wisconsin-Madison, US.

Tsitsi Makombe is a program manager in the West and Central Africa Office of the International Food Policy Research Institute, Dakar, Senegal.

Sohail J. Malik is a visiting fellow in the Development Strategy and Governance Division of the International Food Policy Research Institute, Washington, DC.

Paswel Marenya is a postdoctoral fellow in the Environment and Production Technology Division of the International Food Policy Research Institute, Washington, DC.

Geraldo B. Martha Jr. is a researcher at the Brazilian Agricultural Research Corporation (Embrapa), Brasília.

John McDermott is director of the CGIAR Research Program on Agriculture for Improved Nutrition and Health led by the International Food Policy Research Institute, Washington, DC.

Ruth Meinzen-Dick is a senior research fellow in the Environment and Production Technology Division of the International Food Policy Research Institute, Washington, DC.

Siwa Msangi is a senior research fellow in the Environment and Production Technology Division of the International Food Policy Research Institute, Washington, DC.

Gerald C. Nelson is a senior research fellow in the Environment and Production Technology Division of the International Food Policy Research Institute, Washington, DC.

Alejandro Nin Pratt is a research fellow in the Development Strategy and Governance Division of the International Food Policy Research Institute, Washington, DC.

Ephraim Nkonya is a senior research fellow in the Environment and Production Technology Division of the International Food Policy Research Institute, Washington, DC.

Tolulope Olofinbiyi is a research analyst in the Director General's Office of the International Food Policy Research Institute, Washington, DC.

Steven Were Omamo is director and representative to the African Union and the United Nations Economic Commission for Africa at the World Food Programme, Addis Ababa, Ethiopia.

Robert Paarlberg is Betty F. Johnson Professor of Political Science at Wellesley College in Boston and associate professor at Harvard University, Cambridge, Massachusetts.

Amanda Palazzo is a research analyst in the Environment and Production Technology Division of the International Food Policy Research Institute, Washington, DC.

Rajul Pandya-Lorch is head of the 2020 Vision Initiative at the International Food Policy Research Institute and chief of staff in the Director General's Office, Washington, DC.

Amber Peterman is a research fellow in the Poverty, Health, and Nutrition Division of the International Food Policy Research Institute, Kampala, Uganda.

Prabhu Pingali is deputy director for agricultural development at the Bill & Melinda Gates Foundation, Seattle, Washington.

Beatriz da Silveira Pinheiro is director of the Strategic Studies and Capacity Strengthening Division at the Brazilian Agricultural Research Corporation (Embrapa), Brasília.

Nilam Prasai is a data curator in the Communications Division of the International Food Policy Research Institute, Washington, DC.

Agnes Quisumbing is a senior research fellow in the Poverty, Health, and Nutrition Division of the International Food Policy Research Institute, Washington, DC.

Ana Ramirez is a research analyst in the Director General's Office of the International Food Policy Research Institute, Washington, DC.

Jagdeesh Rao Puppala is executive director of the Foundation for Ecological Security, Anand, India.

Claudia Ringler is deputy director of the Environment and Production Technology Division at the International Food Policy Research Institute, Washington, DC.

Alexander J. Stein is a research coordinator in the Director General's Office of the International Food Policy Research Institute, Washington, DC.

M. S. Swaminathan is chairman of the M. S. Swaminathan Research Foundation, Chennai, India, and a member of Parliament.

Maria Theresa Tenorio is a program analyst in the Environment and Production Technology Division of the International Food Policy Research Institute, Washington, DC.

Peter Timmer is Thomas D. Cabot Professor of Development Studies Emeritus at Harvard University, Cambridge, Massachusetts.

Maximo Torero is director of the Markets, Trade, and Institutions Division of the International Food Policy Research Institute, Washington, DC. He is also the Institute's director for Latin America.

Klaus von Grebmer is a senior research fellow and strategic advisor in the Director General's Office at the International Food Policy Research Institute, Washington, DC.

Stanley Wood is a senior research fellow in the Environment and Production Technology Division of the International Food Policy Research Institute, Washington, DC.

Derek Yach is senior vice president for global health and agriculture policy at PepsiCo, New York.

Sivan Yosef is a project manager in the Director General's Office of the International Food Policy Research Institute, Washington, DC.

Bingxin Yu is a research fellow in the Development Strategy and Governance Division of the International Food Policy Research Institute, Washington, DC.

Made in United States
North Haven, CT
04 November 2026

21744639R00071